Lecture Notes in Control and Information Sciences 432

For further volumes:
http://www.springer.com/series/642

Dmitry Altshuller

Frequency Domain Criteria for Absolute Stability

A Delay-Integral-Quadratic Constraints Approach

 Springer

Author
Dmitry Altshuller
Riverside
California
USA

ISSN 0170-8643   e-ISSN 1610-7411
ISBN 978-1-4471-4233-1   e-ISBN 978-1-4471-4234-8
DOI 10.1007/978-1-4471-4234-8
Springer London Heidelberg New York Dordrecht

Library of Congress Control Number: 2012941049

Printed on acid-free paper

Springer is part of Springer Science+Business Media (www.springer.com)

To Professor Vladimir Andreevich Yakubovich, my lifelong mentor

Preface

The problem of absolute stability has a very long history from its original publication in 1944 by Lurye[1] and Postnikov. Nevertheless, the interest in this problem among control systems theorists is not diminishing. On the contrary, it is increasing. Moreover, some of the new problems, such as robustness and control of systems with uncertainties, can in some sense be viewed as reformulations of this classical problem.

Roughly speaking, the problem is concerned with stability of systems consisting of a linear and a nonlinear block. Only partial information about the latter is given, and stability criteria must involve only the linear block and the known properties of the nonlinear one.

The goal of this book is to bring together some of the most significant results that appeared recently in journals and conference proceedings. Particular attention is paid to the relatively new method, known as delay-integral-quadratic constraints. It turns out that this method sheds some additional insights on a few classical results and also makes it possible to extend them to new classes of systems. Specifically, a number of stability criteria, known previously for autonomous systems, can be extended to time-dependent and, in particular, periodic cases.

As the title of the book suggests, the results are presented in frequency domain, the form in which they tend to naturally arise. In most cases, frequency-domain criteria can be converted to computationally tractable linear matrix inequalities. However, in some cases, inferences concerning system stability can be made directly from the frequency-domain inequalities, especially those that have a certain geometric interpretation, which we discuss in more detail than many other books on the subject.

The book is written in the traditional "theorem-proof" format (except for Chap. 1). However, it is hoped that it could be read by a control systems engineer having a standard background in linear control systems and a certain level of "mathematical maturity." Some of the more technically difficult proofs are given in separate sections at the end of chapters. These sections can be skipped without loss of continuity.

Some of the necessary prerequisites that may be unfamiliar to a control systems engineer are reviewed in the Appendix. The goal of this Appendix is not to give a rigorous theory but rather to "demystify" such concepts as measure, Lebesgue integration, and function spaces. These concepts are used extensively in the formulations of the stability criteria.

---

[1] I believe this is the correct transliteration of the Russian name. Other transliterations that have been used are "Lur'e," "Lurie," and "Lure."

The monograph is organized as follows.

Chap. 1 is a historical survey. While there are many surveys of the problem of absolute stability, greater focus is given to the frequency-domain methods, from the classical Popov criterion, through the development of the method of integral-quadratic constraints. Major milestones are reviewed and some applications are discussed.

Mathematical foundations are laid out in Chap. 2. They include the so-called quadratic criterion for absolute stability and some integral inequalities, the concepts used throughout the book to prove the main results.

Chap. 3 focuses on the so-called stability multipliers. A generalization of the classical result of Zames and Falb is presented. This generalization turns out to be applicable to nonstationary systems. In addition, we discuss a geometric interpretation of stability multipliers, which apparently has not yet been discussed in monographic literature, only in Russian journal publications in the 1980s.

In Chap. 4, the attention is shifted to time-periodic systems. It turns out that stability multipliers, discussed in Chap. 3, take a certain specific form when the nonlinear block is time-periodic. Furthermore, the geometric interpretation discussed in Chap. 3 can also be applicable. The chapter concludes with discussion of some open problems.

This book is an expanded version of my Ph.D. dissertation presented to the Department of Theoretical Cybernetics of St. Petersburg State University (Russia). Many of these ideas were developed under the guidance of my long-time mentor, Professor Vladimir Andreevich Yakubovich. It is to him that I dedicate this work with deep gratitude.

Contents

Chapter 1
A Historical Survey

1.1 Introduction

The problem of absolute stability was formulated by Lurye and Postnikov [94] in 1944 and can be described as follows. Consider the system of equations:

$$\dot{x} = Ax + b\xi, \sigma = c^* x, \xi = \varphi(\sigma, t) . \tag{1.1.1}$$

Here $x, b, c \in \mathbb{R}^m, \xi \in \mathbb{R}, \sigma \in \mathbb{R}$. Usually, the function $\varphi(\sigma, t)$ is assumed to depend only on the variable σ and satisfy the conditions:

$$0 \le \frac{\varphi(\sigma)}{\sigma} \le \kappa, \varphi(0) = 0 . \tag{1.1.2}$$

The double inequality in (1.1.2) will be called the sector condition.

The objective is to find the conditions involving the matrix A as well as the vectors b and c such that the system (1.1.1) is globally asymptotically stable for all functions $\varphi(\sigma)$ satisfying the conditions (1.1.2). This problem is sometimes called the Lurye problem and it has also been associated with the names of Aizerman and Kalman, because of the two conjectures described below.

This problem has a very long history from the original publication by Lurye and Postnikov [94] to the recent monograph by Liao and Yu [83]. Liberzon [87] in his 2006 survey claims to have studied more than 2,000 papers on the subject, of which he listed about a quarter.

Major milestones in the investigation of this problem are marked by the monographs [3, 71, 75, 82, 83, 109, 111, 115, 118, 124, 138, 167] and surveys [25, 56, 84, 87, 98, 110, 119, 142, 160, 162]. Absolute and the related problem of input-output stability also occupy one or more chapters in the books [40, 52, 53, 54, 58-61, 68, 72, 78-81, 130, 139, 143, 144]. None of these three lists is by any means complete. Their length, however, certainly underscores the continued interest in this problem, despite its long history. Furthermore, some of the new problems of robustness and stability of systems with uncertainties can in some sense be viewed as reformulations of the classical problem of absolute stability. Barabanov [23]

D. Altshuller: Frequency Domain Criteria for Absolute Stability, LNCIS 432, pp. 1–24.
springerlink.com   © Springer-Verlag London 2013

gives four possible reformulations of the Aizerman problem and also discusses its relationship with the problem of robustness.

It is futile to even attempt to give a complete history of this problem. The goal of this chapter is far more modest: to present some of the major milestones that are the most relevant to the subject of this book, which is the frequency-domain criteria.

It probably makes the most sense to begin this historical survey with two conjectures associated with the names of Aizerman [2] and Kalman [63]. Aizerman's conjecture states that the absolute stability problem could be solved by replacing a nonlinearity in the form $\varphi(\sigma) = \alpha(\sigma)\sigma$ with a linear gain $\alpha\sigma$, where α is a constant satisfying the double inequality $0 \leq \alpha \leq \kappa$ and using the standard stability criteria for the resulting linear system. Determination of whether this conjecture is true or false became known as the Aizerman problem.

It was proved by Pliss [115] that this conjecture is true for second-order stationary systems, but generally false for systems of order three and higher. Barabanov [23] made a thorough study of the nonstationary third-order case.

However, the matter is not completely settled for systems with delays, the problem described by Rasvan [125] in a chapter of the book *Unsolved Problems in Mathematical Systems and Control Theory*. Rasvan himself proved that the conjecture is true for retarded first-order systems and it has been shown [10] that it is true for retarded second-order systems with a single delay. The word *retarded* means that delays do not involve the highest-order derivative. Three additional classes of time-delay systems, for which the conjecture is true, have been identified in [12].

A second conjecture along the same line was proposed in 1957 by Kalman [63]. His idea was to replace the conditions (1.1.2) with the stronger ones, known as the slope restriction:

$$0 \leq \frac{\varphi(\sigma_1) - \varphi(\sigma_2)}{\sigma_1 - \sigma_2} \leq \kappa, \varphi(0) = 0 . \tag{1.1.3}$$

Stated differently, (1.1.3) is satisfied by monotone Lipschitz nonlinearities.

The assertion that Aizerman's linearization can be used to solve the thus modified absolute stability problem came to be known as the Kalman conjecture. Determination of its truthfulness is known as the Kalman problem. It has a more interesting history than that of Aizerman.

First, it is worth noting that the conjecture is trivially true if local, as opposed to global, stability is considered. Indeed, for the local case it is nothing more than a restatement of the principle of the linearized stability. However, the answer is not so obvious for the global case.

In 1966, Fitts [48] described what he thought was a counterexample. This view was believed to be true until 1987 and was repeated in the well-known monograph

by Narendra and Taylor [109]. Fitts' counterexamples, however, were refuted by
Barabanov [19], who subsequently proved [21] that the Kalman conjecture is true
for all third-order systems and generally false for systems of order four and higher.
For systems with time delays, the Kalman problem is completely unsolved.

In the next section we will describe the attempt to solve the Lurye problem by
using Lyapunov functions. This leads to difficulties, but Lurye himself proposed
a method of resolving them, which leads to linear matrix inequalities. At the time
they were first formulated, there were no effective methods for solving these
inequalities. However, the Kalman-Yakubovich Lemma offered a way to ascer-
tain the existence of a solution. The frequency-domain criteria thus emerged.
At the same time, Popov pioneered another frequency-domain approach, which
led to a stability criterion that now bears his name. Since then, a number of
frequency-domain criteria have been derived, and some of them, especially the
ones involving the so-called Zames-Falb multipliers, will be reviewed in this
chapter.

The integral-quadratic constraints method represents further progress in the frequency-domain approach to stability analysis in that it offers a procedure for deriving stability criteria from the known (or assumed) properties of the system. This
book is about a further development of this method, called the delay-integral-
quadratic constraints approach, which makes it possible to considerably extend the
scope of its applicability.

1.2 Resolving Equations and Matrix Inequalities

Let us try to solve the Lurye problem by using the following Lyapunov function
candidate:

$$V(x,\sigma) = x^*Px + \beta \int_0^{\sigma} \varphi(\sigma)d\sigma , \qquad (1.2.1)$$

where P is a positive definite matrix.

Calculating the derivative of this function along the trajectories of the system
(1.1.1), we find

$$\dot{V}(x,\xi) = x^*(PA + A^*P)x + \left(Pb + \frac{1}{2}A^*c \right)^* x\xi + \beta c^*b\xi^2$$

$$= 2(Px + \beta c\xi)^*(Ax + b\xi).$$

Clearly, the right-hand side vanishes when $Ax + b\xi = 0$. Therefore, the derivative
of the Lyapunov function candidate $V(x,\sigma)$ cannot be negative definite. This difficulty is pointed out by a number of authors, including [3, 75, 83].

In order to resolve this difficulty, Lurye[1] used the following approach. Add to and subtract from the expression for $\dot{V}(x,\xi)$ the function $\xi(c^*x - \kappa^{-1}\xi)$. Since $\sigma = c^*x$ and $\xi = \varphi(\sigma)$, this function is positive definite in view of the conditions (1.1.2).

The expression for $\dot{V}(x,\xi)$ becomes

$$\dot{V}(x,\xi) = -S(x,\xi) - \xi(c^*x - \kappa^{-1}\xi) \, ,$$

with the function $S(x,\xi)$ given by

$$S(x,\xi) = -x^*(PA + A^*P)x - 2\xi H^*x + \gamma^2\xi^2 \, ,$$

where

$$H = PB + \frac{1}{2}(c + \beta A^*c);$$
$$\gamma^2 = \kappa^{-1} - \beta c^*b.$$

Hence, $\dot{V}(x,\xi)$ is negative definite if $S(x,\xi)$ is positive definite.

Let Q be an arbitrary positive definite matrix. Then the requirement for the quadratic form $S(x,\xi)$ to be positive definite yields the following equation for the unknown matrix P:

$$\begin{bmatrix} -(PA + A^*P) & Pb + \frac{1}{2}(c + \beta A^*c) \\ \left(Pb + \frac{1}{2}(c + \beta A^*c)\right)^* & \gamma^2 \end{bmatrix} = Q. \tag{1.2.2}$$

This matrix equation (a system of equations in the original work by Lurye) is called the system of resolving equations. The parameter β can be chosen arbitrarily. Clearly, the system is absolutely stable if there exists a matrix P satisfying (1.2.2).

The Lurye method played a dominant role in the research on the absolute stability problem until the emergence of the frequency-domain approach in 1960. Many important results can be found in the book by Letov [82]. The method was later named "the S-procedure" by Aizerman and Gantmacher [3] because of the

----

[1] Lurye developed the method for $\kappa^{-1} = 0$. It was then extended to the general case by Aizerman and Gantmacher [3].

notation $S(x, \xi)$. Some theoretical aspects of the S-procedure are discussed in the Yakubovich's paper [159].

Clearly, the equation (1.2.2) can instead be written as a linear matrix inequality (LMI) for the same unknown matrix P. There is a considerable amount of literature on LMIs, including the renowned monograph by Boyd and his coworkers [33]. Some major contributions were made by Kamenetskii [65, 66] as well as by Pyatnitskii and his coworkers [67, 103-105, 121, 122].

Development of effective solution methods for LMIs led some researchers to express stability criteria in this form. Examples of this approach are the books by Korenevskii [71], Liao and Yu [83], and Niculescu [111], the paper by Grujic [55], and the above-mentioned work by Kamenetskii [65].

A more computationally involved approach, also based on LMIs, uses the so-called inners, which are certain determinants. This method is described in depth in the monograph by Jury [62] and the survey paper by Liberzon [87].

1.3 The Emergence of the Frequency-Domain Methods

At the time when Lurye developed his method of resolving equations, there were no known effective methods of solving these equations or LMIs. However, it is not actually necessary to find such solution. All that is needed is to prove the existence of such a solution. This can often be accomplished using the frequency-domain methods that have taken a dominant place in the theory of absolute stability. They involve the transfer function of the linear block defined for the system (1.1.1) by the equation

$$W(s) = c^*(sI - A)^{-1}b .$$

The advantage of the frequency-domain stability criteria is that many of them have a convenient geometric interpretation, similar in spirit to the Nyquist criterion for linear systems. In fact, the circle criterion uses the Nyquist plot, and the Popov criterion uses a modified form of the Nyquist plot.

In this section we shall describe in the historical sequence the classical result of V. M. Popov, the contributions of V. A. Yakubovich, and the subsequent development by other researchers. The presentation will be informal, without rigorous statements and proofs of theorems.

1.3.1 The Popov Criterion and Its Modifications

Historically, Popov's result was the first of the frequency-domain criteria. It was first published in 1961 and Popov's original paper [117] was subsequently reprinted in the volume *Control Theory: Twenty-Five Seminal Papers* [30].

It is stated as follows. The system (1.1.1) with nonlinearity satisfying the conditions (1.1.2) is absolutely stable if there exists a real number θ, such that for all the real values of ω

$$\kappa^{-1} + \mathrm{Re}\big[(1 + i\omega\theta)W(i\omega)\big] > 0 . \tag{1.3.1}$$

This is an extremely powerful result. It makes it possible for many nonlinear control problems to be solved analytically. The monograph by Rasvan [124] discusses a number of applications. He also discusses three approaches in using the criterion.

The first approach is the direct analysis of the inequality (1.3.1) for the given transfer function. Typically, this can be done only for systems of low order. Let us illustrate this approach by proving that the Aizerman conjecture is true for retarded second-order systems [12]. A simpler case is treated in [10].

Consider the second-order retarded differential equation:

$$\ddot{\sigma}(t) + \alpha_1 \dot{\sigma}(t) + \sum_{j=1}^{m} \beta_j \sigma(t - \tau_j) + \varphi(\sigma) = 0 \tag{1.3.2}$$

with the nonlinearity $\varphi(\sigma)$ satisfying the sector condition (1.1.2).

It follows immediately from the results of Pontryagin [116] (see also [46, p. 176]) that the necessary and sufficient condition for the stability of the corresponding linear system

$$\ddot{\sigma}(t)(t) + \alpha_1 \dot{\sigma}(t) + \alpha \sigma + \sum_{j=1}^{m} \beta_j \sigma(t - \tau_j) = 0 \tag{1.3.3}$$

is that the following three inequalities hold:

$$\sum_{j=1}^{m} |\beta_j| < \alpha, \ 2\sum_{j=1}^{m} |\beta_j| < \alpha_1^2, \ \alpha_1 > 0. \tag{1.3.4}$$

The first of these inequalities implies that in order for the zero solution of (1.3.2) to be globally asymptotically stable, the nonlinearity $\varphi(\sigma)$ must satisfy the inequality

$$\varphi(\sigma) > \sigma \sum_{j=1}^{m} |\beta_j|.$$

This suggests defining a new nonlinearity:

$$\phi(\sigma) = \varphi(\sigma) - \sigma \sum_{j=1}^{m} |\beta_j|.$$

The transfer function of the linear part then becomes

$$W(s) = \left[s^2 + \alpha_1 s + \sum_{j=1}^{m} |\beta_j| + \sum_{j=1}^{m} \beta_j e^{-\tau_j s} \right]^{-1}.$$

Expansion of (1.3.1) yields

$$\sum_{j=1}^{m} |\beta_j| + (\alpha_1 \theta - 1) \omega^2 - \theta \omega \sum_{j=1}^{m} \beta_j \sin \omega \tau_j + \sum_{j=1}^{m} \beta_j \cos \omega \tau_j \geq 0.$$

We can replace this inequality with

$$\sum_{j=1}^{m}\Big[\big|\beta_j\big|+\gamma_j\left(\alpha_1\theta-1\right)\omega^2-\theta\omega\beta_j\sin\omega\tau_j+\beta_j\cos\omega\tau_j\Big]\geq 0 \qquad (1.3.5)$$

where $\gamma_j, j=1\ldots m$ is a set of positive real numbers such that $\sum_{j=1}^{m}\gamma_j=1$. Clearly, (1.3.5) holds if each summand is nonnegative. Using a well-known trigonometric identity, we can rewrite this in the form

$$\sum_{j=1}^{m}\Big[\big|\beta_j\big|+\gamma_j\left(\alpha_1\theta-1\right)\omega^2-\theta\omega\beta_j\sin\omega\tau_j+\beta_j\cos\omega\tau_j\Big]\geq 0. \qquad (1.3.6)$$

It is easy to demonstrate that (1.3.6) holds if θ is chosen to satisfy the double inequality:

$$\frac{\gamma_j\alpha_1}{\big|\beta_j\big|}-\sqrt{\frac{\gamma_j\left(\gamma_j\alpha_1^2-2\big|\beta_j\big|\right)}{\big|\beta_j\big|^2}}<\theta<\frac{\gamma_j\alpha_1}{\big|\beta_j\big|}+\sqrt{\frac{\gamma_j\beta_1^2-2\big|\beta_j\big|}{\big|\beta_j\big|^2}}. \qquad (1.3.7)$$

Furthermore, we set

$$\gamma_j=\big|\beta_j\big|\Big/\sum_{j=1}^{m}\big|\beta_j\big|. \qquad (1.3.8)$$

The inequality (1.3.7) then becomes

$$\frac{\alpha_1}{\sum_{j=1}^{m}\big|\beta_j\big|}-\sqrt{\frac{\alpha_1^2-2\sum_{j=1}^{m}\big|\beta_j\big|}{\left(\sum_{j=1}^{m}\big|\beta_j\big|\right)^2}}<\theta<\frac{\alpha_1}{\sum_{j=1}^{m}\big|\beta_j\big|}+\sqrt{\frac{\alpha_1^2-2\sum_{j=1}^{m}\big|\beta_j\big|}{\left(\sum_{j=1}^{m}\big|\beta_j\big|\right)^2}}.$$

In light of the second of the inequalities (1.3.4), the radicands are positive. Therefore, stability of the zero solution of the linear equation (1.3.3) implies that a constant θ can be found to satisfy the Popov inequality, which means that the zero solution of (1.3.2) is globally asymptotically stable. The Aizerman conjecture for this type of system is proved.

If, in addition, $\beta_1=0$, we obtain the well-known fact that the Aizerman conjecture is true for second-order systems.

The second approach is the geometric analysis. It involves drawing in the complex plane the parametric curve $\left(\operatorname{Re}W(i\omega),\omega\operatorname{Im}W(i\omega)\right)$ for $\omega\in[0;+\infty)$. The inequality (1.3.1) is satisfied if and only if this plot (commonly called the Popov plot) lies entirely to the right of the straight line given by $x=\theta y-\kappa^{-1}$ with x representing the real axis and y representing the imaginary axis. The parameter θ can be chosen in the most advantageous way.

Let us illustrate this idea with a numerical example. Consider the system with a linear block given by the transfer function

$$W(s) = \frac{40}{s(s+1)(s^2+0.8+16)} \ . \tag{1.3.9}$$

Set the parameters as follows:

$$\theta = 1; \kappa^{-1} = 2.9 \ . \tag{1.3.10}$$

The plot, shown in Fig. 1.1, indicates that (1.3.1) holds. Hence, the system is absolutely stable.

The third approach is, essentially, the numerical implementation of the geometric analysis. The inequality (1.3.1) can be rewritten in the form

$$\kappa^{-1} > \min_{\theta} \max_{\omega} \left[-\mathrm{Re}\,W(i\omega) + \theta\omega\,\mathrm{Im}\,W(i\omega) \right]. \tag{1.3.11}$$

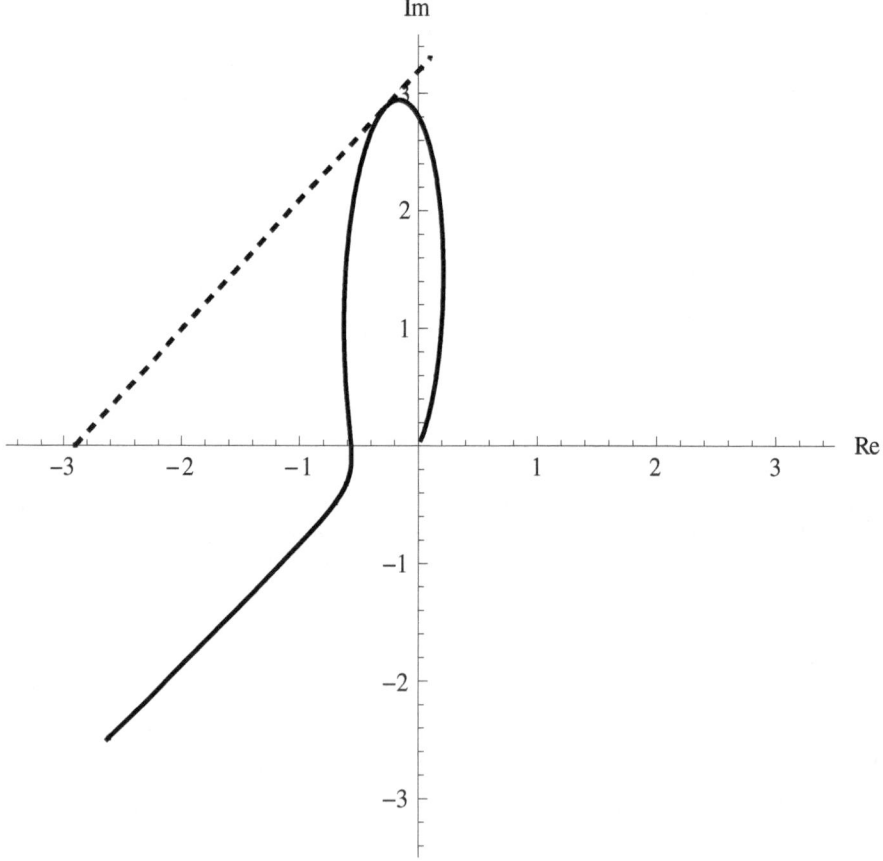

Fig. 1.1 Popov plot for the transfer function given by (1.3.9) and the parameters given by (1.3.10)

The value of the right-hand side of (1.3.11) can then be evaluated numerically (or analytically, if possible).

An interesting modification of the Popov criterion, convenient for using the third approach, was proposed by Liao and Yu [83]. For a given rational frequency response $W(i\omega)$ they write its real and imaginary parts in the form:

$$\operatorname{Re} W(i\omega) = \frac{A(\omega)}{E(\omega)} \; ; \; \operatorname{Im} W(i\omega) = \frac{B(\omega)}{E(\omega)} \, .$$

Then they define the polynomials:

$$A(\omega) + \kappa^{-1} E(\omega) = c_{2n} + c_{2n-1}\omega + \ldots + c_0 \omega^{2n}$$
$$B(\omega) = b_{2n} + b_{2n-1}\omega + \ldots + b_0 \omega^{2n},$$

and three constants

$$\rho_1 = 1 + \max_{1 \le j \le 2n} \left| \frac{c_j}{c_0} \right| \; ; \; \rho_2 = 1 + \max_{1 \le j \le 2n} \left| \frac{b_j}{b_0} \right| \; ; \; \rho = \max(\rho_1, \rho_2) \, .$$

They then proceed to prove that the system is absolutely stable if one of the following two conditions holds:

1) $b_0 < 0$ (or $b_0 = 0, b_1 < 0$ and there exists a constant $\theta \ge 0$ such that (1.3.1) holds for $\omega \in [0; \rho]$.

2) $b_0 > 0$ (or $b_0 = 0, b_1 > 0$ and there exists a constant $\theta \le 0$ such that (1.3.1) holds for $\omega \in [0; \rho]$.

The Popov plot is also used in a so-called parabola criterion due to Bergen and Sapiro [31]. They prove that the system is absolutely stable if there exist constants θ_1, θ_2, and θ_3 such that the parabola

$$x^2 + \frac{\theta_1 + 2\theta_2}{\theta_2(\theta_1 + \theta_2)} x - \frac{\theta_1 \theta_3}{\theta_2(\theta_1 + \theta_2)} y + \frac{1}{\theta_2(\theta_1 + \theta_2)} = 0$$

defined in the plane of complex variables $z = x + iy$, does not contain either the origin or any part of the Popov plot.

Another improvement for the Popov criterion worth mentioning is its robust version proved by Dahleh et al. [42]. Essentially, they have combined the Popov criterion with the well-known Kharitonov's robustness theorem and obtained a stability condition applicable to a wide class of transfer functions.

The question of whether the Popov criterion is a necessary, as well as suffi-cient, condition for absolute stability remained open until 1973, when Pyatnitskii [120] proved that there exist absolutely stable systems, for which the inequality (1.3.1) does not hold for any real value of the constant θ. That said, there are sys-tems for which the Popov criterion is a necessary as well as sufficient condition for absolute stability. Some of them are identified in the paper [76] as well as the monographs [3, 83, 115].

However, it follows from the Kalman-Yakubovich Lemma, described in the next subsection, that the Popov criterion completely characterizes the class of stationary systems, for which the question of absolute stability can be resolved by using the Lyapunov function in the form (1.2.1). Another frequency-domain criterion that can be obtained by using this type of the Lyapunov function was derived by Rekasius and Rowland [126]. Their result is applicable to nonstationary systems but imposes some additional requirements on the nonlinearity. Leonov [77] also proposed an extension of the Popov-type frequency condition to nonstationary systems.

1.3.2 Kalman-Yakubovich Lemma and the Circle Criterion

Yakubovich met Popov's breakthrough with one of his own, and his paper [147] has also taken its rightful place in the volume *Control Theory: Twenty-Five Se-minal Papers* [30]. His result later came to be known as the Kalman-Yakubovich Lemma, due to subsequent work by Kalman [64]. Some authors call it the Positive Real Lemma. More detailed discussion can be found in Yakubovich's papers [149-151].

There are several ways of stating this remarkable result. Here we follow the monograph by Leonov and Smirnova [81].

Let (A, b) be a completely controllable pair and let $\mathcal{G}(x, \xi)$ be a quadratic form. The Kalman-Yakubovich Lemma states that a Hermitian matrix $P = P^*$, such that for all $\forall x \in \mathbb{R}^m$, $\forall \xi \in \mathbb{R}$: $2x^* P(Ax + b\xi) + \mathcal{G}(x, \xi) \leq 0$, exists if and on-ly if the inequality

$$\mathrm{Re}\,\mathcal{G}((i\omega I - A)^{-1} b\xi, \xi) < 0$$

holds for all real values of ω and all complex values of ξ. Furthermore, if the matrix A is Hurwitz, then the matrix P is positive definite.

One of the immediate consequences of this theorem is the celebrated circle criterion. Consider the system (1.1.1) with the function $\varphi(\sigma, t)$ satisfying the conditions

$$\kappa_1 \leq \frac{\varphi(\sigma, t)}{\sigma} \leq \kappa_2, \varphi(0, t) \equiv 0 \,. \tag{1.3.12}$$

Clearly, the double inequality is equivalent to

$$\mathcal{G}(x,\xi) \triangleq (\xi - \kappa_1 c^* x)(\xi - \kappa_2 c^* x) \geq 0 . \qquad (1.3.13)$$

Let us consider $V(x) = x^* P x$, where P is a symmetric matrix, as a Lyapunov function candidate. Computing its derivative we find

$$\dot{V}(x) = 2x^* P(Ax + b\xi) .$$

By using the S-procedure we find that $\dot{V}(x)$ is negative definite if the quadratic form $2x^* P(Ax + b\xi) + \mathcal{G}(x,\xi)$ is negative definite. Application of the Kalman-Yakubovich Lemma yields the inequality:

$$\operatorname{Re}\left\{ [1 + \kappa_1 W(i\omega)]^* [1 + \kappa_2 W(i\omega)] \right\} > 0 .$$

The last inequality is called the circle criterion due to the fact that the equation

$$\operatorname{Re}\left\{ [1 + \kappa_1 z]^* [1 + \kappa_2 z] \right\} = 0 .$$

defines a circle in the complex plane with the center on the real axis and passing through two points: $\left(-\kappa_1^{-1}, 0\right)$ and $\left(-\kappa_2^{-1}, 0\right)$. If $\kappa_1 = 0$ (any sector condition can be reduced to this case by sector rotation), the circle degenerates into vertical straight line, and the Nyquist plot of $W(s)$ must lie to the right of this line.

The circle criterion can be formally obtained from the Popov criterion by taking $\theta = 0$. However, the Popov criterion is applicable only to systems with nonlinearities depending only on the variable σ, while the circle criterion does not require such restriction. This type of tradeoff occurs very frequently in the theory of absolute stability.

Some results, similar to the circle criterion, were proved by Bongiorno [36] (for linear nonstationary systems) and Narendra and Goldwyn [107] (for nonlinear stationary systems). Yakubovich [158] later extended the circle criterion to proving instability of certain systems. Megrestki [100] derived a result that he called a multiloop generalization of the circle criterion.

It can be shown that one of the consequences of the Kalman-Yakubovich Lemma is that the circle criterion covers all the systems, for which the question of absolute stability can be resolved by using a quadratic form as a Lyapunov function. Another consequence is that the Popov criterion covers all the systems for which this question can be resolved by using a quadratic form plus integral of the nonlinearity as a Lyapunov function candidate, as mentioned in the previous subsection.

The latter fact led to yet another extension of the Popov criterion, proved in [169]. It is concerned with local, as opposed to global, stability. First, let us note that for the Popov criterion the quadratic form $\mathcal{G}(x,\xi)$ is given by

$$\mathcal{G}(x,\xi) = \xi(\kappa c^* x - \xi) + \vartheta c^* (Ax + b\xi)\xi.$$

Assume that the frequency condition in the Popov criterion is satisfied and let P be the Hermitian matrix whose existence is guaranteed by the Kalman-Yakubovich Lemma. Denote

$$\rho = c^* P^{-1} c \; ; \; \Phi(\sigma) = \rho^{-1}\sigma^2 + \vartheta \int_0^\sigma \varphi(\sigma)d\sigma \; ; \; V(x) = x^* Px + \vartheta \int_0^{c^* x} \varphi(\sigma)d\sigma \, .$$

Suppose that the sector condition $\varphi(\sigma)[\sigma - \kappa^{-1}\varphi(\sigma)] \geq 0$ is satisfied for $\sigma_1 \leq \sigma \leq \sigma_2$. Then the domain of attraction of the equilibrium point $x = 0$ is given by $V(x) \leq \min[\Phi(\sigma_1), \Phi(\sigma_2)]$.

The Kalman-Yakubovich Lemma is a deep and far reaching theorem. In addition to the theory of absolute stability, it proved useful in the investigation of adaptive control systems and in solving some optimal control problems. Interested readers may consult the 2006 survey by Gusev and Likhtarnikov [56] for more details. It has also found its use in the analysis and synthesis of electrical networks [16].

1.3.3 Subsequent Development

The results of Popov and Yakubovich established two approaches to the problem of absolute stability: Popov used the method of a priori integral estimates, while Yakubovich used Lyapunov functions and the S-procedure, which led to the theorem now known as the Kalman-Yakubovich Lemma.

Popov's method proved extremely useful in the investigation of systems with integral equations and distributed parameters (see, for example, the book by Leonov, Ponomarenko, and Smirnova [79]). Yakubovich's method was subsequently extended to systems with multiple equilibrium states, which also include systems with cylindrical state space and, in particular, phase-locked loops [167].

Most of the work on the absolute stability problem was focused on the Kalman problem, i.e., systems with nonlinearities satisfying the slope restriction condition

$$0 \leq \frac{\varphi(\sigma_1) - \varphi(\sigma_2)}{\sigma_1 - \sigma_2} \leq \kappa, \varphi(0) = 0 \, . \tag{1.3.14}$$

Many results proved in the 1960s and 1970s have the form:

$$\text{Re}\{Z(i\omega)[\kappa^{-1} + W(i\omega)]\} > 0, \tag{1.3.15}$$

where $Z(s)$ is a function called the stability multiplier. Clearly, both the circle ($Z(s) \equiv 1$) and the Popov ($Z(s) = 1 + \theta s$) criteria are special cases of this form. Here are some of the most important results.

In 1962 Yakubovich [148] proved the criterion of the form (1.3.15) with

$$Z(s) = a + bs + cs^2, a \geq 0, c \geq 0, a + |b| + c > 0.$$

We will prove this result, often referred to as the Yakubovich criterion, in Subsect. 3.2.3 using the method of quadratic constraints, discussed later. Here we note that it was used by Barabanov [21] to prove that the Kalman conjecture is true for third-order systems.

In 1965, Brockett and Willems [34, 35] obtained the criterion of the form (1.3.15) with

$$Z(s) = a_0 s + \sum_{j=1}^{N} \frac{a_j (s + b_j)}{c_j s + b_j}, \ a_j \geq 0, \ b_j \geq 0, \ c_j \in (0;1).$$

Similar forms were obtained by Narendra and his coworkers Cho [105] and Neuman [108].

Another criterion with an interesting geometric interpretation was proved in 1968 by Cho and Narendra [39]: The system is absolutely stable if there exists a constant θ, such that for all positive real values of ω

$$\mathrm{Re}\left\{(1 + i\theta)\left[\kappa^{-1} + W(i\omega)\right]\right\} > 0.$$

This result came to be known as the off-axis circle criterion. A similar result was derived by Voronov [141].

Application of the off-axis circle criterion involves drawing in the complex plane the standard Nyquist plot of the transfer function $W(i\omega)$ and then drawing a straight line of the slope θ and intersecting the real axis at the point κ^{-1}. This is similar to the geometric interpretation of the Popov criterion except that the Nyquist plot is used directly instead of its modified form.

Let us revisit the numerical example considered earlier as an application of the Popov criterion:

$$W(s) = \frac{40}{s(s+1)(s^2 + 0.8 + 16)} . \tag{1.3.16}$$

Set the parameters as follows:

$$\theta = 2.1; \ \kappa^{-1} = 0.85 . \tag{1.3.17}$$

From the plot shown in Fig. 1.2 we can conclude that the system is absolutely stable. Note that this criterion produces a less conservative stability margin compared to the Popov criterion. The tradeoff is that the off-axis circle criterion is applicable only to a narrower class of systems, namely, the slope-restricted nonlinearities.

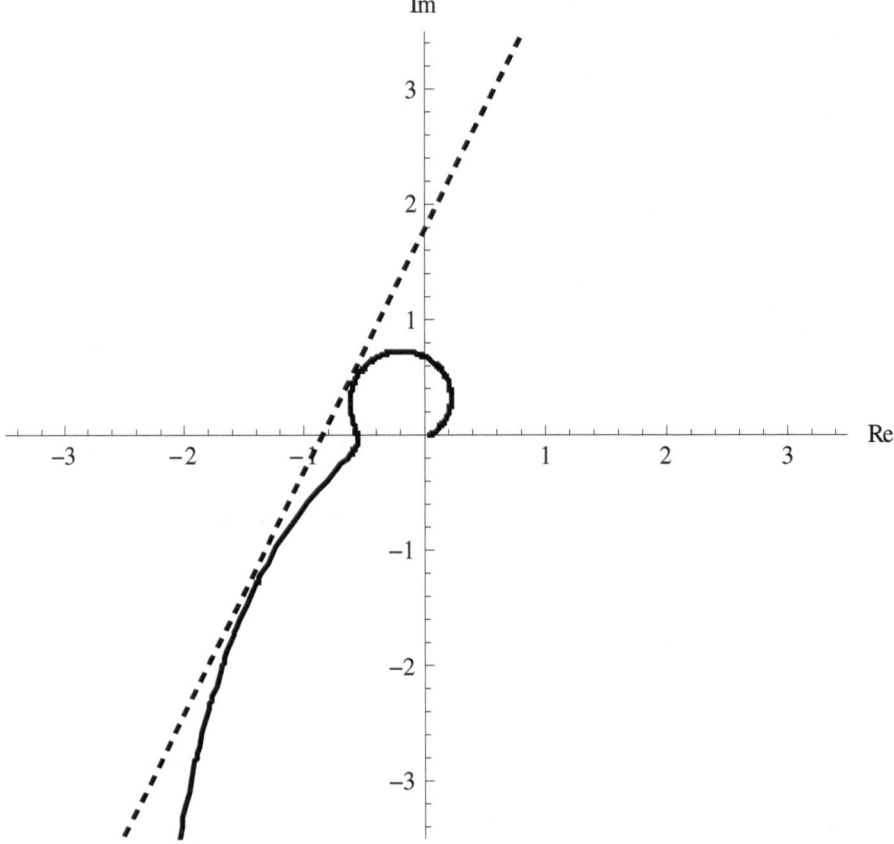

Fig. 1.2 Off-axis circle plot for the transfer function given by (1.3.16) and the parameters given by (1.3.17)

A very general form of the stability multiplier was derived by Barabanov [19]. The system is absolutely stable if there exist a real constant θ and a real function $y(\tau)$ with the L_1-norm less than 1 such that $y(\tau) \leq 0$ unless the nonlinearity is odd and for all real values of τ

$$\mathrm{Re}\left\{\left[\kappa^{-1} + W(i\omega)\right]\left[1 + i\omega\theta - \int_{-\infty}^{+\infty} y(\tau)e^{i\omega\tau}d\tau\right]\right\} > 0. \qquad (1.3.18)$$

In the same paper [19], Barabanov also proved a more constructive stability criterion, namely, that (1.3.18) holds with $y(\tau) \leq 0$ if there exist sets of numbers

$\{\alpha_n, \beta_n\}_{n=1}^p$, $\{\gamma_n, \delta_n\}_{n=1}^r$, θ_0, θ_1, such that $\varepsilon > 0, 0 \le \alpha_1 < \beta_1 < \ldots < \alpha_p < \beta_p$, $\max\{\theta_1 \beta_p, \theta_1 \delta_r\}$, and for all $\omega \ge 0$

$$\mathrm{Re}\left\{\left[W(i\omega) + \kappa^{-1}\right](1 + i\omega\theta_1)\prod_{n-1}^{p}\frac{i\omega + \alpha_n}{i\omega + \beta_n}\prod_{n-1}^{r}\frac{i\omega + \gamma_n}{i\omega + \delta_n}\right\} \ge \varepsilon > 0 .$$

Stated differently, the stability multiplier is given in the form

$$Z(s) = (1 + \theta_1 s)\prod_{n-1}^{p}\frac{s + \alpha_n}{s + \beta_n}\prod_{n-1}^{r}\frac{s + \gamma_n}{s + \delta_n} .$$

An immediate consequence of this result is the frequency condition of the off-axis circle criterion. Similar forms of the stability multiplier were obtained by O'Shea [112, 113] and Yakubovich [150].

An interesting general result concerning the forms of the multipliers was proved by Venkatesh [137]. A general algebraic procedure for constructing multipliers was proposed by Sundareshan and Thathachar [135].

It is worth pointing out that some stability criteria do not have the multiplier form. One such result has been proved by Dewey and Jury [45]. Barabanov [22] derived some results for a class of nonstationary nonlinearities. We will see some others in Chap. 4.

Barkin (and Zelentsovskii) [27-29] proposed yet another approach to derivation of the frequency-domain criteria. It is based on the so-called power transformation of vectors and matrices. A concise description can be found in the survey [110].

Many important results are collected in the monographs by Popov [118] and Narendra and Taylor [109], both of which appeared in 1973. Since then, the only book completely devoted to the subject of absolute stability was written by Liao and Yu [83]. This book focuses mainly on the S-procedure and some special type of control systems, giving little attention to the frequency-domain methods.

1.4 The Problem of Input-Output Stability

The problem of input-output stability arises naturally from the frequency-domain treatment of nonlinear systems, which can, in some sense, be viewed as an analogue of the classical control theory of linear systems. Indeed, as we have seen in the previous section, several stability criteria rely on the Nyquist plot of the frequency response of the linear block.

The problem was formulated by Zames in two papers [170, 171], both of which are also reprinted in [30]. It is concerned with the feedback interconnection shown in Fig. 1.3.

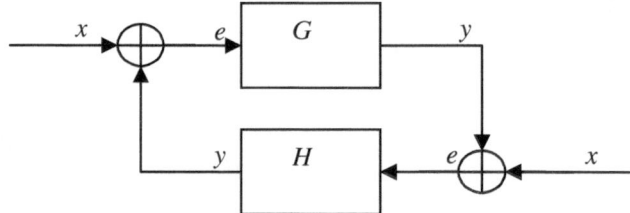

Fig. 1.3 Feedback interconnection

Here G is usually a linear operator, which may be expressed as a transfer function and H can be either a linear or a nonlinear operator. The problem can be stated as follows: Given that the signals x_1 and x_2 belong to an L^p space, is it true that signals e_1 and e_2 belong to the same L^p space? This property is called the L^p-stability. For L^∞ spaces the problem is often referred to as bounded-input-bounded-output (BIBO) stability.

Among other results, proved by Zames in [171], are the circle and the Popov criteria, both of which are sufficient for the L^2-stability, which is the most commonly studied.

The most general result was proved in 1967 by Zames and Falb [173], which has the familiar multiplier form (1.3.15) with

$$Z(s) = 1 + \sum_{j=1}^{\infty} a_j e^{b_j s} + \int_{-\infty}^{+\infty} e^{-st} \vartheta(t) dt .$$

The following inequalities must also be satisfied:

$$\int_{-\infty}^{+\infty} \vartheta(t) dt + \sum_{j=1}^{\infty} a_j > -1, a_j < 0, \vartheta(t) \leq 0 .$$

For this reason stability multipliers are often called the Zames-Falb multipliers. Subsequently, Zames, together with his colleagues Falb and Freedman [47, 50], extended this concept to a more abstract formulation. Generally speaking, the problem of input-output stability lent itself very naturally to the functional analysis approach.

The problem of input-output stability is treated extensively in the monograph by Willems [145], the textbook by Desoer and Vidyasagar [44], and in chapters of the books [43, 53, 54, 67, 139]. As we will see in Sect. 2.1, this problem and that of absolute stability have become essentially the same and will be considered together.

1.5 The Method of Integral-Quadratic Constraints

The origins of the method of integral-quadratic constraints lie in the S-procedure, which involves adding to, and subtracting from, the expression for the derivative of the Lyapunov function a certain quadratic form $\mathcal{F}(x,\xi)$, known to be positive definite. For example, an immediate consequence of the sector condition is the inequality $(\xi - \kappa_1 \sigma)(\xi - \kappa_2 \sigma) \geq 0$. The idea of the method is that certain properties of the nonlinearity imply that a certain inequality of the form $\mathcal{F}(\sigma(t), \xi(t)) \geq 0$ or, more generally, $\mathcal{F}(\xi(t), \sigma(t), \dot{\sigma}(t)) \geq 0$ holds for all values of t. This condition is called a local quadratic constraint.

Absolute stability of systems satisfying this type of condition was investigated in the renowned paper by Yakubovich [153]. He subsequently extended it to discrete systems [155, 156]. It turned out that several criteria known at that time could be derived using this approach by finding the appropriate quadratic constraints. Conditions for absolute instability can also be derived using this approach [157, 158].[2]

An important generalization of a local quadratic constraint is an integral-quadratic constraint: There exist a sequence $t_k \to \infty$ and a positive constant γ such that

$$\int_0^{t_k} \mathcal{F}\big(\xi(t), \sigma(t), \dot{\sigma}(t)\big) dt + \gamma \geq 0, \quad \gamma \geq 0. \tag{1.5.1}$$

Constraints of this form are said to be in time domain in order to distinguish them from the ones in frequency domain discussed later. They first appear in the above-cited paper by Yakubovich [153]. Note that if the system satisfies more than one constraint, we can take a linear combination of the quadratic forms.

The quadratic form in either local- or integral-quadratic constraints can be extended to a Hermitian form $\tilde{\mathcal{F}}(\tilde{\xi}, i\omega) \triangleq \mathcal{F}\big(\tilde{\xi}, -W(i\omega)\tilde{\xi}, -i\omega W(i\omega)\tilde{\xi}\big)$ of the variable $\tilde{\xi}$.

The early statement of the so-called quadratic criterion for absolute stability requires that thus defined Hermitian form $\tilde{\mathcal{F}}(\tilde{\xi}, i\omega)$ be negative definite. There are other conditions that must be met as well, but we are not going to discuss them here. This statement can be easily extended to the case of more than one constraint. One early application of this result was to derive stability conditions for systems with a hysteresis-type nonlinearity [26].

Any integral-quadratic expression in time domain can be, using the Plancherel theorem, rewritten in frequency domain:

$$\int_0^{+\infty} \mathcal{F}(\xi(t), \sigma(t)) dt = \int_{-\infty}^{+\infty} \mathcal{F}(\tilde{\xi}(\omega), \tilde{\sigma}(\omega)) d\omega, \tag{1.5.2}$$

---

[2] The title of the paper [157] is incorrectly translated as "absolute stability" instead of "absolute instability"!

which leads to a constraint:

$$\int_{-\infty}^{+\infty} \mathcal{F}(\tilde{\xi}(\omega), \tilde{\sigma}(\omega)) d\omega \geq 0 .$$

This can be done only if the integral in the left-hand side of (1.5.2) converges, i.e., only for some of the solutions of the original system. This is an important limitation discussed below.

The matrix of the quadratic form in (1.5.2) has thus far been assumed to be constant. This concept can be generalized by defining constraints in frequency domain using an arbitrary Hermitian matrix $\Pi(i\omega)$:

$$\int_{-\infty}^{+\infty} \begin{bmatrix} \tilde{\xi}(\omega) \\ \tilde{\sigma}(\omega) \end{bmatrix}^* \Pi(i\omega) \begin{bmatrix} \tilde{\xi}(\omega) \\ \tilde{\sigma}(\omega) \end{bmatrix} d\omega \geq 0 .$$

This is the key concept in the paper by Megrestki and Rantzer [100]. However, they impose some limitations, such as that nonlinearity must be bounded in order to assure convergence of the integrals. When all the appropriate limitations are met, the requirement for stability is written in the form

$$\begin{bmatrix} I \\ W(i\omega) \end{bmatrix}^* \Pi(i\omega) \begin{bmatrix} I \\ W(i\omega) \end{bmatrix} \leq -\varepsilon I .$$

So, indeed, the constraints in frequency domain are a natural extension of the constraints in time domain.

The limitations imposed by Megrestki and Rantzer [100] were weakened by Yakubovich in a series of papers [163, 165, 166]. However, this still leaves open the question of obtaining the constraints. Megrestki and Rantzer [100] give several interesting examples of doing so.

In this book we are going to consider constraints in time domain, but also involving values of variables at some earlier point in time (i.e., time delays). The advantage of doing so is that, as we shall see, these constraints arise naturally from certain properties of input and output signals.

1.6   Two Applications

1.6.1   Flying Vehicle

Consider the axisymmetric flying vehicle rotating around its longitudinal axis. The equations describing the dynamics of such a vehicle are well known [131, 134, 175]:

$$\begin{aligned}
\dot{\alpha} &= \omega_z - c_\alpha \alpha + c_{\alpha M}\, \omega_x \beta + c_\delta \delta_y + Lg/V^2 \\
\dot{\beta} &= \omega_y - c_\alpha \beta - c_{\alpha M}\, \omega_x \alpha + c_\delta \delta_z \\
\dot{\omega}_z &= J \omega_x \omega_y + m_\alpha \alpha + m_{\alpha M}\, \omega_x \beta - m_\omega \omega_z - m_\delta \delta_y \\
\dot{\omega}_y &= -J \omega_x \omega_z + m_\alpha \beta - m_{\alpha M}\, \omega_x \alpha - m_\omega \omega_y - m_\delta \delta_z .
\end{aligned} \tag{1.6.1}$$

The dot in these equations denotes differentiation with respect to the dimensionless time τ.

The state variables in these equations are the angles of attack and yaw α and β, respectively, and the dimensionless angular velocities ω_z and ω_y of transverse oscillations of the axis of symmetry. The control inputs are δ_y and δ_z. The parameters are as follows:

ω_x – the dimensionless angular velocity of rotation around the axis of symmetry (roll);

V – the velocity of the center of mass of the vehicle;

L – the length of the vehicle;

J – the ratio of the axial and transverse moments of inertia;

$c_\alpha, c_{\alpha M}$ – reduced aerodynamic coefficients of the lift and the Magnus [32] force, respectively;

$m_\alpha, m_\omega, m_{\alpha M}$ – reduced aerodynamic coefficients of the restoring torque, damping torque, and the Magnus force, respectively;

c_δ, m_δ – reduced aerodynamic coefficients of the control force and torque.

The objective is to determine conditions for stability of the balanced flight, that is when $\omega_y = \omega_z = 0$. The values of the other state variables during such flight will be denoted by the superscript 0.

The first criterion for stability of the balanced flight was proposed by Soshnikov and Fedorova [134]. They assumed that V and ω_x are, indeed, constant parameters that do not change in the course of the flight. They showed that under these assumptions, the following inequality is a necessary and sufficient condition for stability:

$$\left(\omega_x^0\right)^2 \left[m_{\alpha m}^{\;2} + m_{\alpha m}\left(c_\alpha - m_\alpha\right)\left(J - c_{\alpha M}\right) - c_\alpha m_\omega \left(J - c_{\alpha M}\right)^2 \right]$$
$$< \left(c_\alpha + m_\alpha\right)^2 \left(c_\alpha m_\omega - m_\alpha\right).$$

However, as pointed out by Zhermolenko and Lokshin [175], both V and ω_x may change during the flight due to variations in the applied thrust or random wind gusts. Therefore, it is more realistic to assume that they are unknown functions of time, bounded by inequalities:

$$V_1 \leq V(\tau) \leq V_2;$$
$$\omega_x^- \leq \omega_x(\tau) \leq \omega_x^+.$$

The problem, therefore, can be considered an absolute stability problem.

Zhermolenko and Lokshin used one of the results from Yakubovich's paper [154] to obtain the following sufficient condition for absolute stability:

$$\left(\omega_x^+\right)^2 < -\frac{\left(c_\alpha + m_\omega\right)\left[m_\alpha + \left(c_\alpha - m_\omega\right)^2\big/4\right]}{4\left(1 - m_\alpha\right)\left(J^2 + c_{\alpha M}^{\;2} + m_{\alpha M}^{\;2}\right)}. \tag{1.6.2}$$

It is worth noting that physically reasonable assumptions about system parameters are

$$m_\alpha < 0,\; c_\alpha > 0,\; m_\omega > 0.$$

Therefore, the necessary condition for (1.6.2) to hold is

$$\left|m_\alpha\right| > \left(c_\alpha - m_\omega\right)^2\big/4.$$

Further contribution to the problem was made by Skorodinskii [131] who, using the method from [121], proved that the condition (1.6.2) is not only sufficient but also necessary.

Liberzon [85] considered a somewhat different version of the problem. He assumed that the velocity $V(\tau)$ can be treated as a random perturbation of a constant V^0:

$$V(\tau) = V^0 + v(\tau).$$

Furthermore, it can be assumed that

$$\left|v(\tau)\right| \le a.$$

Moreover, Liberzon also pointed out that, as shown in Benton's paper [32], the angular velocity $\omega_x(\tau)$ is proportional to the velocity $V(\tau)$ and the proportionality coefficient can be a function of time, which we can assume to be bounded by the inequality

$$0 \le k(\tau) \le K.$$

Therefore,

$$\omega_x(\tau) = k(\tau)\left[V^0 + v(\tau)\right].$$

Moreover, the state variables in (1.6.1) can be replaced with their deviations. The terms Lg/V^2 as well as those containing the control inputs δ_y and δ_z can then be omitted. With this in mind, the system (1.6.1) becomes

$$
\begin{aligned}
\dot{\alpha} &= \omega_z - c_\alpha \alpha + c_{\alpha M} V^0 k(\tau)\beta + c_{\alpha M} k(\tau) v(\tau)\beta \\
\dot{\beta} &= \omega_y - c_\alpha \beta - c_{\alpha M} c_{\alpha M} V^0 k(\tau)\alpha - c_{\alpha M} k(\tau) v(\tau)\alpha \\
\dot{\omega}_z &= JV^0 k(\tau)\omega_y + Jk(\tau) v(\tau)\omega_y + m_\alpha \alpha \\
&\quad + m_{\alpha M} V^0 k(\tau)\beta + m_{\alpha M} k(\tau) v(\tau)\beta - m_\omega \omega_z \\
\dot{\omega}_y &= -JV^0 k(\tau)\omega_z - JV^0 k(\tau) v(\tau)\omega_z + m_\alpha \beta \\
&\quad - m_{\alpha M} V^0 k(\tau)\alpha - m_{\alpha M} k(\tau) v(\tau)\alpha - m_\omega \omega_y.
\end{aligned}
\tag{1.6.3}
$$

The sufficient condition for absolute stability obtained by Liberzon, as improved in his subsequent paper [86], using the method of inners [62] has the form of two inequalities, both of which must hold:

$$
\left(c_\alpha - m_\omega\right)^2 + 4m_\alpha > 2K^2 \left(V^0 + a\right)^2 \left(J^2 + c_{\alpha M}{}^2\right)
$$

and

$$
\begin{aligned}
&\left[K^2 a^2 \left(J^2 + c_{\alpha M}{}^2\right) + \left(c_\alpha + m_\omega\right)^2 + 2\left(c_\alpha m_\omega - m_\alpha\right)\right]^2 \\
&+ 4K^2 a^2 \left[J\left(c_\alpha J + m_{\alpha M}\right) + c_{\alpha M}\left(c_{\alpha M} m_\omega + m_{\alpha M}\right)\right]\left(c_\alpha + m_\omega\right) \\
&> 2K^2 \left(V^0 + a\right)^2 \left[\left(J^2 + c_{\alpha M}{}^2\right)\left(c_\alpha + m_\omega\right)^2 + 2\left(c_\alpha J + m_{\alpha M}\right)^2\right] \\
&+ 4\left[c_{\alpha M} K^2 \left(V^0 + a\right)^2 J + m_\alpha\right]^2 - 8c_\alpha m_\omega m_\alpha.
\end{aligned}
$$

The problem presently open is to refine the last condition or to determine if it is also necessary.

1.6.2 Hydraulic Regulator

Consider a hydraulic regulator shown schematically in Fig. 1.4. This device translates the displacement of the spool C, denoted by ζ, to the displacement of the piston B, denoted by η. Time constants of both the sleeve A and the spool B are assumed to be negligible.

The mathematical model of this system, which we now describe, was constructed in a short paper by Kheifetz and Gelig [69]. With the assumption that the

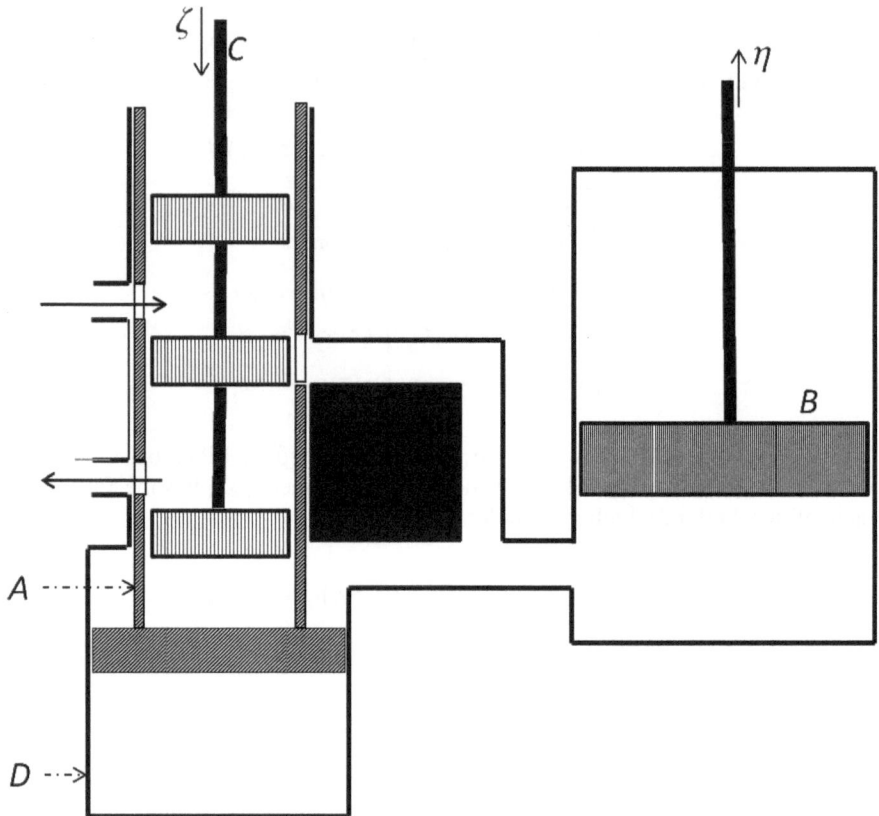

Fig. 1.4 Regulator Schematic

normalized (relative to the pressure under the piston) friction force of the spool B is much less than the normalized friction force between the moving sleeve A and the walls of the chamber D, the two displacements are related by the equation:

$$\eta = \zeta + \varphi(\dot{\zeta}).$$

The function φ represents the difference between the normalized friction forces. It is assumed to satisfy the following conditions:

$$\operatorname{sgn} \varphi(\dot{\zeta}) = \operatorname{sgn} \dot{\zeta};$$

$$\lim_{\dot{\zeta} \to 0+} \varphi(\dot{\zeta}) \triangleq \varphi_+ > \varphi_- \triangleq \lim_{\dot{\zeta} \to 0-} \varphi(\dot{\zeta}).$$

Furthermore, it is assumed that there exists a number $\delta > 0$ such that the following two inequalities hold:

$$\varphi(\dot{\zeta}) > \varphi_+ - \delta\dot{\zeta} \quad \text{if } \dot{\zeta} > 0;$$
$$\varphi(\dot{\zeta}) < \varphi_- - \delta\dot{\zeta} \quad \text{if } \dot{\zeta} < 0.$$

The simplified dynamics of the system consisting of the just described this regulator, the servomotor and the turbine can be described by the following equations [69, 164, 165]:

$$T_1 \frac{dx_1}{dt} = -x_1 + x_3 - \xi(t)$$
$$T_2 \frac{dx_2}{dt} = x_1 - x_2$$
$$T_3 \frac{dx_3}{dt} = -x_2$$
$$\sigma(t) = x_2(t)$$
$$\xi(t) = \varphi(\sigma(t)),$$

where x_1, x_2, and x_3 are suitably normalized η, $\dot{\zeta}$, and ζ, respectively.

All state variables are real scalars; $T_1>0$, $T_2>0$, and $T_3>0$ are the time constants of the servomotor, steam volume, and the turbine itself. The function $\varphi(\sigma)$ is used to describe dry friction and has a discontinuity of the first kind at the point $y = 0$. It satisfies the conditions $\sigma\varphi(\sigma) \geq 0$; $\varphi(0+0) = 1$; $\varphi(0-0) = -1$.

The solution of this system of equations (due to discontinuity) is understood in the sense described in the book by Yakubovich, Leonov, and Gelig [167]. The characteristic polynomial of the linear block of the system is

$$\Delta(s) = (1+T_1 s)(1+T_2 s)T_3 s + 1.$$

The stability condition for this linear block is

$$T_1 T_2 < (T_1 + T_2)T_3. \tag{1.6.4}$$

The nonlinearity satisfies the quadratic constraints of the form (1.5.1) for the following two quadratic forms:

$$\mathcal{F}_1(\xi,\sigma,\dot{\sigma}) = \xi\sigma; \quad \mathcal{F}_2(\xi,\sigma,\dot{\sigma}) = \xi\dot{\sigma}$$

with

$$\gamma_2 = \int_0^{\sigma[0]} \varphi(\sigma)d\sigma \,.$$

The transfer function between the input $\xi(t)$ and the output $\sigma(t)$ is given by

$$W(s) = \frac{T_3 s}{\Delta(s)} \,.$$

The frequency condition described in Sect. 1.5 leads to the conclusion that the inequality (1.6.4) is a necessary and sufficient condition for the stability of this system.

Chapter 2
Foundations

2.1 Modern Formulation of the Absolute Stability Problem

The classical problem of absolute stability is concerned with the system:

$$\dot{x} = Ax + B\xi \,, \sigma = Cx \,, \xi = \varphi(\sigma,t) \,. \tag{2.1.1}$$

Here $\sigma(t) \in \mathbb{R}^m; \xi(t) \in \mathbb{R}^p; A$, B, and C are constant matrices of appropriate dimensions. Without significant loss of generality, we can assume that the matrix A is Hurwitz stable.

We are going to consider systems defined in a more general form, similar to the one used in the input-output stability problem, described in Sect. 1.4. Specifically, the linear block will be given by a Volterra integral equation:

$$\sigma(t) = \alpha(t) + K_0\xi(t) + \int_0^t K(t-s)\xi(s)ds \,. \tag{2.1.2}$$

The nonlinear block will be represented by the relation:

$$[\sigma(\cdot), \xi(\cdot)] \in \mathsf{N} \,. \tag{2.1.3}$$

The system (2.1.1) can be written in this form by setting $K(t) = Ce^{At}B$, $\alpha(t) = Ce^{At}x(0)$, and $K_0 = 0$. The set N is defined by the equation $\xi = \varphi(\sigma,t)$.

Clearly, if the nonlinear block is represented as $\xi = \varphi(\sigma,t)$, then the system can be reduced to a nonlinear Volterra integral equation. Existence theory for this type

D. Altshuller: Frequency Domain Criteria for Absolute Stability, LNCIS 432, pp. 25–41.
springerlink.com

of equations is well developed (see, for example, the classic book by Miller [102]), and we will not be concerned with this question.

We are going to assume that the linear block satisfies the regularity conditions defined as follows.

Definition 2.1. *The linear block* (2.1.2) *satisfies the regularity conditions if*

$$|\alpha(\cdot)| \in L^2(0;+\infty) \cap L^\infty(0;+\infty) \qquad (2.1.4)$$

and there exist positive constants C *and* β *such that*

$$|K(t)| \le Ce^{-\beta t}. \qquad (2.1.5)$$

Clearly, these conditions are satisfied for the system (2.1.1) provided that the matrix A is Hurwitz.

We are going to consider only locally square-integrable signals. In other words, it will be assumed that $|\sigma(\cdot)| \in L^2_{loc}(0,\infty)$ and $|\xi(\cdot)| \in L^2_{loc}(0,\infty)$, where $L^2_{loc}(0,\infty)$ denotes a set of functions, square-integrable on any interval $[0;t]$. Stated differently, $L^2_{loc}(0,\infty)$ is an extended space $L^2(0;+\infty)$ or the space of functions with globally square-integrable truncations.

For the purposes of this chapter, we define the set N as follows: For any $\xi(\cdot), \sigma(\cdot) \in L^2_{loc}(0,\infty)$, there exists a sequence $t_k \to \infty$, possibly dependent on $\sigma(\cdot)$ and $\xi(\cdot)$ such that for some numbers $\gamma_j \ge 0$

$$\forall \tau \in \mathbb{T}_j \subseteq \mathbb{R}_+ : \int_0^{t_k} \mathcal{F}_j\big(\sigma(t),\xi(t),\sigma(t-\tau),\xi(t-\tau)\big)dt + \gamma_j \ge 0, j=1,2,...N. \quad (2.1.6)$$

Here $\mathbb{T}_j$ are certain subsets of nonnegative real numbers (to be defined later) and $\mathcal{F}_j\big(\sigma_1,\xi_1,\sigma_2,\xi_2\big): \mathbb{R}^m \times \mathbb{R}^m \times \mathbb{R}^m \times \mathbb{R}^m \to \mathbb{R}$ are quadratic forms. For the purposes of stating the quadratic criterion for absolute stability, these forms are assumed to be given. If they do not depend on either σ_2 or ξ_2, then the conditions (2.1.6) reduce to constraints in time domain [153]. The inequalities (2.1.6) are called the delay-integral-quadratic constraints.

Let us define the following sets:

$$Z_{loc} = \left\{ z(\cdot) = [\sigma(\cdot), \xi(\cdot)] : |z(\cdot)| \in L^2_{loc}(0; \infty) \right\},$$

$$Z = \left\{ z(\cdot) = [\sigma(\cdot), \xi(\cdot)] : |z(\cdot)| \in L_{loc}(0; \infty) \right\},$$

$$L_{loc} = \left\{ z(\cdot) = [\sigma(\cdot), \xi(\cdot)] : \text{ equation (2.1.2) holds} \right\},$$

$$L_0 = \left\{ z(\cdot) = [\sigma(\cdot), \xi(\cdot)] : \text{ equation (2.1.2) holds with } \alpha(t) \equiv 0 \right\},$$

$$L = L_{loc} \cap Z,$$

$$M_\gamma = \left\{ z(\cdot) = [\sigma(\cdot), \xi(\cdot)] : \text{ equation (2.1.6) holds for some } t_k \to \infty \right\}.$$

Let $\gamma_j[z(\cdot)]$ denote the set γ_j in (2.1.6) corresponding to the process $z(\cdot)$.

The elements of the set Z_{loc} are called processes, and the elements of the set Z are called stable processes. In addition, we will need the set $M^\infty_{\gamma_j[z(\cdot)]} \subset Z$ of stable processes satisfying

$$\forall \tau \in \mathbb{T}_j \subseteq \mathbb{R}_+ : \int_0^\infty \mathcal{F}_j\big(\sigma(t), \xi(t), \sigma(t-\tau), \xi(t-\tau)\big) dt + \gamma_j \geq 0, j = 1, 2, \ldots N.$$

Definition 2.2. *The system* (2.1.2), (2.1.3) *is absolutely stable if all processes* $L \cap N$ *are stable and there exists a constant* λ*, same for all processes, such that*

$$\|z(\cdot)\| \leq \lambda \left(\|\alpha(\cdot)\| + \sum_j \gamma_j[z(\cdot)] \right).$$

This is the definition used in most of the recent papers on absolute stability, including [4-8, 11-15, 163-166]. It bears a strong resemblance to the definition of the L^2-stability discussed in Sect. 1.4, which explains similarities in some of the known results.

Our objective will be to find conditions that must be imposed on the linear block in order to assure absolute stability of the system.

2.2 Quadratic Criterion

The quadratic criterion for absolute stability is the "cornerstone" for all the results presented in this book. Historically, it was proved by Yakubovich for differential equations and then for integral equations. There are several ways of presenting it. Here we follow the articles [15] and [166]. Other formulations can be found in several papers by Yakubovich [153, 160, 163-165]. An abstract form, applicable to systems defined on Banach spaces, was proved in [161].

Most of the results in the book will be formulated in terms of the frequency response of the linear block defined with the help of the Fourier transform of its kernel:

$$W(i\omega) = -K_0 - \tilde{K}(i\omega) = -K_0 - \int_0^t K(t)e^{-i\omega t}\,dt. \tag{2.2.1}$$

Each of the quadratic forms $\mathcal{F}_j\left(\sigma_1,\xi_1,\sigma_2,\xi_2\right)$ can be extended to a Hermitian form $\mathcal{F}_j\left(\tilde{\sigma}_1,\tilde{\xi}_1,\tilde{\sigma}_2,\tilde{\xi}_2\right)$ with $\tilde{\sigma}_1,\tilde{\xi}_1,\tilde{\sigma}_2,\tilde{\xi}_2 \in \mathbb{C}^m$. Define the Hermitian matrix $\Pi_j(\omega,\tau)$ by the equation:

$$\tilde{\xi}*\Pi_j(i\omega,\tau)\tilde{\xi} = \mathcal{F}_j\left(-W(i\omega)\tilde{\xi},\tilde{\xi},-W(i\omega)\tilde{\xi}e^{-i\omega\tau},\tilde{\xi}e^{-i\omega\tau}\right).$$

Let the matrix F_j of each of the quadratic forms $\mathcal{F}_j\left(\sigma_1,\xi_1,\sigma_2,\xi_2\right)$ be written as

$$F_j = \begin{bmatrix} F_{11j} & F_{12j} & F_{13j} & F_{14j} \\ F_{12j}^* & F_{22j} & F_{23j} & F_{24j} \\ F_{13j}^* & F_{23j}^* & F_{33j} & F_{34j} \\ F_{14j}^* & F_{24j}^* & F_{34j}^* & F_{44j} \end{bmatrix}.$$

Then the matrix $\Pi_j(\omega,\tau)$ is given by

$$\begin{aligned}
\Pi_j(\omega,\tau) &= W*(i\omega)\left(F_{11j} + F_{13j}e^{-i\omega\tau} + F_{13j}^*e^{i\omega\tau} + F_{33}\right)W(i\omega) \\
&\quad -W*(i\omega)\left(F_{12j} + F_{14j}e^{-i\omega\tau} + F_{23j}^*e^{i\omega\tau} + F_{34j}\right) \\
&\quad -\left(F_{12j}^* + F_{14j}^*e^{i\omega\tau} + F_{23j}e^{-i\omega\tau} + F_{34j}^*\right)W(i\omega) \\
&\quad +F_{22j} + F_{24j}e^{-i\omega\tau} + F_{24j}^*e^{i\omega\tau} + F_{44} \\
&= W*(i\omega)\left(F_{11j} + F_{13j}e^{-i\omega\tau} + F_{13j}^*e^{i\omega\tau} + F_{33}\right)W(i\omega) \\
&\quad +F_{22j} + F_{24j}e^{-i\omega\tau} + F_{24j}^*e^{i\omega\tau} + F_{44} \\
&\quad -2\,\mathrm{Re}\left[\left(F_{12j} + F_{14j}e^{i\omega\tau} + F_{23j}^*e^{-i\omega\tau} + F_{34j}\right)W(i\omega)\right].
\end{aligned}$$

Suppose further that there exists a constant matrix Ξ, same for all the matrices F_j such that each one of these matrices satisfies the following additional requirements:

$$F_{13j} = 0;$$
$$F_{11j} + F_{33j} = 0;$$
$$F_{12j} + F_{34j} = \Xi\left(F_{22j} + F_{44j}\right).$$

In addition, assume that one of the following pairs of conditions is met:

$$F_{14j} = F_{23j} = \Xi F_{24j}^* / 2;$$
$$F_{14j} = 0, F_{23j} = \Xi F_{24j}^* / 2; \qquad (2.2.2)$$
$$F_{23j} = 0, F_{14j} = \Xi F_{24j}^* / 2.$$

Then the matrix $\Pi_j(\omega, \tau)$ becomes

$$\Pi_j(\omega, \tau) = \mathrm{Re}\left\{ \left(F_{12j} + F_{14j}e^{i\omega\tau} + F_{23j}^* e^{-i\omega\tau} + F_{34j} \right)\left[\Xi + W(i\omega) \right] \right\}. \qquad (2.2.3)$$

We say that the frequency condition (FC) is satisfied if there exist: 1) a constant $\varepsilon > 0$, and 2) a set of nonnegative finite measures μ_j on $\mathbb{R}_+$, such that $\mu_j(\mathbb{R}_+ \setminus \mathbb{T}) = 0$ and for all real values of ω

$$\sum_{j=1}^{N} \int_0^{+\infty} \Pi_j(\omega, \tau) d\mu_j(\tau) \le -\varepsilon I_m. \qquad (2.2.4)$$

Note that the second factor in the right-hand side of (2.2.3) does not depend on the quadratic constraints. Therefore, it can be factored out of both the integral and the sum in the right-hand side of (2.2.4). It should be clear that not every set of quadratic constraints will yield the frequency condition in this form.

There is another way to formulate the frequency condition. First, note that the set $\mathbb{R}_+ \setminus \mathbb{T}$, if nonempty, is a union of a countable (possibly, finite) set of open intervals. Indeed, let $z(t)$ be an arbitrary process and let $\tau_0 \in \mathbb{R}_+ \setminus \mathbb{T}$. Then for any $t > 0$ and any $\gamma_j > 0$, we have

$$\int_0^t \mathcal{F}_j\left(\sigma(t), \xi(t), \sigma(t - \tau_0), \xi(t - \tau_0) \right) dt + \gamma_j < 0.$$

The integral in the left-hand side is a continuous function of τ_0. Therefore, there is a neighborhood of the number τ_0, such that for all numbers τ in this neighborhood

$$\int_0^t \mathcal{F}_j\left(\sigma(t), \xi(t), \sigma(t - \tau), \xi(t - \tau) \right) dt + \gamma_j < 0.$$

This implies that the set $\mathbb{R}_+ \setminus \mathbb{T}$ is open and, hence, is a union of a countable collection of open intervals.

Recall that every nondecreasing function generates a nonnegative measure. Therefore, the frequency condition can be stated in terms of nondecreasing functions $\vartheta_j(\tau)$, constant on each interval making up the set $\mathbb{R}_+ \setminus \mathbb{T}$. Conversely, every nonnegative measure can be generated by a nondecreasing function. The integral in (2.2.4) can, therefore, be understood in the sense of Lebesgue-Stieltjes. This is the approach used in [7]. It makes some of the proofs more cumbersome.

The following lemma relates the frequency condition with the inequality in the definition of the absolute stability.

Lemma 2.3. *If the frequency condition (2.2.4) is met, then there exists a constant* $\lambda > 0$ *such that for any process* $z(\cdot) \in \mathsf{L} \cap \mathsf{M}_\gamma^\infty$

$$\|z(\cdot)\|^2 \le \lambda \left(\|\alpha(\cdot)\|^2 + \sum \gamma_j [z(\cdot)] \right). \tag{2.2.5}$$

The proof of this lemma is given in Sect. 2.4.

An immediate consequence of Lemma 2.3 is that the system is absolutely stable if it does not have any unstable processes and the frequency condition holds. The first of these properties can be assured by requiring that the system is minimally stable – a concept that we now define. However, first we need to define the notion of a stable continuation of a process.

Definition 2.4. *A stable continuation of a process* $z(\cdot)$ *in* $\mathsf{M}_{\gamma_j[z(\cdot)]}^\infty$ *is a sequence of processes* $z_k(\cdot) \in \mathsf{L} \cap \mathsf{M}_{\gamma_j[z(\cdot)]}^\infty$ *such that* $z_k(t) = z(t)$ *for* $0 \le t \le t_k$ *with* $t_k \to \infty$.

Definition 2.5. *The system (2.2), (2.3) is minimally stable if every process* $z(\cdot) \in \mathsf{L}_{loc} \cap \mathsf{N}$ *has a stable continuation in* $\mathsf{M}_{\gamma_j[z(\cdot)]}^\infty$.

The following theorem will be referred to as the quadratic criterion for absolute stability.

Theorem 2.6. *Suppose that the system (2.1.2), (2.1.3) is minimally stable and the FC (2.11) holds. Then this system is absolutely stable.*

Proof. Let $z(\cdot) \in \mathsf{L}_{loc} \cap \mathsf{N}$ be a process and let $z_k(\cdot)$ be its stable continuation in $\mathsf{M}_{\gamma_j[z(\cdot)]}^\infty$. By Lemma 2.3

$$\|z(\cdot)\|^2 \le \lambda \left(\|\alpha(\cdot)\|^2 + \sum_j \gamma_j [z(\cdot)] \right).$$

Since $z_k(t) = z(t)$ for $0 \le t \le t_k$, we have

$$\int_0^{t_k} |z(t)|^2 \, dt \le \|z_k(\cdot)\|.$$

In the limit as $t_k \to \infty$, we obtain that $|z(\cdot)| \in L^2[0; +\infty)$ and

$$\|z(\cdot)\|^2 \le \lambda \left(\|\alpha(\cdot)\|^2 + \sum_j \gamma_j [z(\cdot)] \right).$$

Therefore, the system is absolutely stable by Definition 2.2.

Thus, in order to apply the quadratic criterion, we need to establish that a given system is minimally stable and derive the frequency condition from the delay-integral-quadratic constraints that the system is known to satisfy. The first of these tasks can often be accomplished with the help of the following lemma.

Lemma 2.7. *Suppose that the linear block (2.1.2) satisfies the regularity conditions, the frequency condition (2.2.4) is met, and $|z(\cdot)| \in L^\infty[0; +\infty)$. Then $|z(\cdot)| \in L^2[0; +\infty)$.*

Proof of this lemma is given in Sect. 2.5.

The property described by this lemma is called dichotomy, which, stated differently, means that any process in the system under consideration is either stable or unbounded. It is an important property that has been studied on its own (see, for example, [17, 18, 41, 79]). For our purposes, its usefulness will be in establishing the property of minimal stability. For the case of quadratic constraints without delays, the result was proved by Yakubovich [153] (see also [40]).

Now let us introduce the concept of a bounded continuation of a process, which will be useful in the application of this theorem.

Definition 2.8. *A bounded continuation of a process $z(\cdot)$ in $\mathbf{M}_\gamma$ is a sequence of processes $z_k(\cdot) \in \mathsf{L}_{\mathrm{loc}} \cap \mathbf{M}_\gamma$ such that $|z_k(\bullet)| \in L^\infty[0, +\infty]$ and $z_k(t) = z(t)$ for $0 \le t < t_k$ with $t_k \to \infty$.*

The following theorem will be used throughout the book to prove the minimal stability of a given system.

Theorem 2.9. *Suppose that the linear block (2.1.2) satisfies the regularity conditions and the frequency condition (2.2.4) is met. Then any bounded continuation of a process $z(\cdot)$ in $\mathbf{M}_\gamma$ is a stable continuation of this process in $\mathbf{M}^\infty_{\gamma[z(\cdot)]}$. If every* *process* $z(\cdot) \in \mathsf{L}_{loc} \cap \mathsf{N}$ *has a bounded continuation in* $\mathsf{M}^{\infty}_{\mathcal{H}z(\cdot)]}$, *the system is minimally stable.*

This theorem is an immediate consequence of Lemma 2.7.

2.3 Two Integral Inequalities

In this section we address the second aspect of applying the quadratic criterion – the derivation of the delay-integral-quadratic constraints. To this end, we shall prove two integral inequalities satisfied by functions of certain types.

The following lemma is an extension of the known result of Willems and Gruber [146]. Some other, more specialized, versions of it were proved in [4-7, 11, 14, 15, 73, 74, 128].

Lemma 2.10. *Assume the following:*

1) *The function* $G : \mathbb{R}^m \times \mathbb{R}_+ \to \mathbb{R}$ *is continuous in each argument. Let*
    $g(x,t) = \nabla_x G(x,t)$;

2) $G(x,t) \ge 0$ *for all* x *and* t;

3) $G(0,t) \equiv 0$;

4) *There exists a constant T, such that for all* t *and* y, $G(y,t-T) \ge G(y,t)$;

5) *There exists a function* $H : \mathbb{R}^m \times \mathbb{R}^m \to \mathbb{R}$, *such that for all values of* x, y, *and* t,

$$G(y,t) - G(x,t) + (\zeta x - y)g(x,t) \ge -H(x, g(x,t)) - H(y, g(y,t)). \quad (2.3.1)$$

Then for any real number b *and any measurable function* x(t), *such that* $x(t) \equiv 0$ *for* $t < 0$, *the following inequality holds:*

$$\int_0^b \{ g(x(t),t) [\zeta x(t) - x(t-T)] + H(x(t), g(x,t),t)) \} dt$$

$$+ \int_0^b H(x(t-T), g(x(t-T),t)) dt \ge 0. \quad (2.3.2)$$

If, in addition, the function g(x,t) *is odd in* x, *then for any real number* b *and any measurable function* x(t), *such that* $x(t) \equiv 0$ *for* $t < 0$, *the following inequality holds:*

$$\int_0^b \left\{ g(x(t),t)[\zeta x(t) + x(t-T)] + H(x(t), g(x,t),t)) \right\} dt$$

$$+ \int_0^b H(x(t-T), g(x(t-T),t)) dt \geq 0. \tag{2.3.3}$$

Proof. Using $G(y,t-T) \geq G(y,t)$, we can replace (2.3.1) with

$$G(y,t-T) - G(x,t) + (\zeta x - y) g(x,t) \geq -H(x, g(x,t) - H(y, g(y,t)). \tag{2.3.4}$$

Now we prove that (2.3.4) implies (2.3.2). To this end, we add to each side of (2.3.2) the expression

$$\int_0^b [G(x(t-T),t-T) - G(x(t),t)] dt .$$

This yields

$$\int_0^b \left\{ g(x(t),t)[\zeta x(t) - x(t-T)] + H(x(t), g(x,t),t)) + H(x(t-T), g(x(t-T),t)) \right\} dt$$

$$+ \int_0^b [G(x(t-T),t-T) - G(x(t),t)] dt$$

$$+ \int_0^b [G(x(t),t) - G(x(t-T),t-T)] dt$$

$$= \int_0^b \left\{ G(x(t-T),t-T) - G(x(t),t) + g(x(t),t)[\zeta x(t) - x(t-T)] \right\} dt$$

$$+ \int_0^b [H(x(t), g(x,t),t)) + H(x(t-T), g(x(t-T),t))] dt$$

$$+ \int_0^b [G(x(t),t) - G(x(t-T),t-T)] dt.$$

Substitution $s = t-T$ yields, keeping in mind that $x(t) \equiv 0$ for $t < 0$:

$$\int_0^b [G(x(t),t) - G(x(t-T),t-T)] dt = \int_0^b G(x(s),s) ds - \int_0^{b-T} G(x(s),s) ds$$

$$= \int_{b-T}^T G(x(s),s) ds.$$

Therefore,

$$\int_0^b \{g(x(t),t)[\zeta x(t) - x(t-T)] + H(x(t),g(x,t),t)) + H(x(t-T),g(x(t-T),t))\}dt$$

$$= \int_0^b \{G(x(t-T),t-T) - G(x(t),t) + g(x(t),t)[\zeta x(t) - x(t-T)]\}dt$$

$$+ \int_0^b [H(x(t),g(x,t),t)) + H(x(t-T),g(x(t-T),t))]dt$$

$$+ \int_{b-T}^T G(x(s),s)ds.$$

The last integral is nonnegative because $G(x,t) \geq 0$ for all x and t. By setting $x=x(t)$, $y=x(t-T)$ and applying (2.3.1), we conclude that the sum of the first two integrals is also nonnegative. Therefore, the entire expression is nonnegative, which proves the inequality (2.3.2).

Suppose now that the function $g(x,t)$ is odd in x. Then, the function $G(x,t)$ is even in x, and we can replace $G(y,t)$ with $G(-y,t)$ in (2.3.1). The inequality (2.3.3) is proved by repeating the above arguments.

Other forms of this lemma have been proved in the earlier papers [4-6, 11, 14]. Safonov and Kulkarni [128] also proved a special form of this result.

Note that the condition 3) of the lemma is automatically satisfied by functions of one variable and by functions periodic in t if T is the period (the inequality degenerates into an identity). Furthermore, it is satisfied for all values of T by functions that are nonincreasing in t for positive x and nondecreasing in t for negative x. Such functions will sometimes be called semimonotone.

Let us exhibit a less trivial example of a function satisfying this condition. Let $g(x,t)=g_1(x)\,g_2(t)$ and assume that $g_1'(x) \leq 1$ for all x. Define the function $g_2(t)$ as follows. Let $f(t)$ be a decreasing function for which there exist positive constants c_1 and c_2 such that

$$c_1 \leq f(t) \leq c_2, \quad \lim_{t \to \infty} f(t) = c_1.$$

Let $g_2(t)$ be an arbitrary function, such that $f(t+s) \leq g_2(t) \leq f(t)$ for some positive number s. It is not difficult to see that the function $g(x,t)$ thus defined satisfies the condition 3) of the lemma.

Figure 2.1 shows a numerical example of such functions for $T = 2$ with

$$s = 1.9$$

$$f(t) = e^{-t} + 1$$

$$g_2(t) = f(t)\sin^2 6t + f(t+s)\cos^2 6t.$$

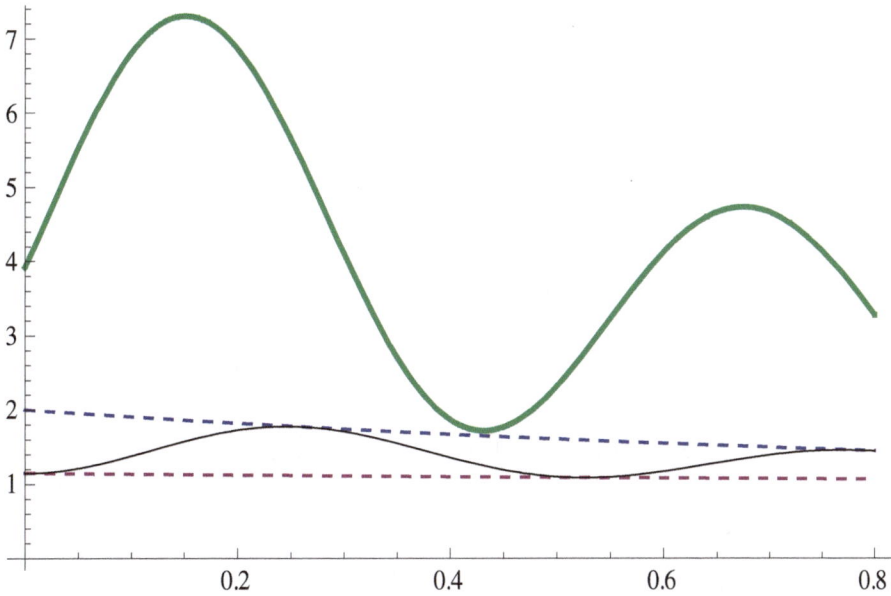

Fig. 2.1 Functions $f(t)$, $f(t+s)$, shown with dashed lines, and $g_2(t)$, $g_2(t-T)$, shown with thin and thick solid lines, respectively

2.4 Proof of Lemma 2.3

The proof of Lemma 2.3 relies, in turn, on the following lemma, which is, in fact, an abstract form of the quadratic criterion, stated in Hilbert space and proved by Yakubovich [166].

Lemma 2.11. *Let Z be a Hilbert space, L_0 its (possibly incomplete) subspace, $z_0 \in Z$, and $L=L_0+z_0$. Let $\mathcal{A}(z)=z{*}Fz$, where F is a linear bounded self-adjoint operator, be a quadratic form. Define $N_\gamma = \{z : \mathcal{F}(z)+\gamma \geq 0\}$, where γ is a positive constant. Assume that there exists a positive constant δ, such that*

$$\forall y \in L_0 \quad \mathcal{F}(y) \leq -\delta \|y\|^2 . \tag{2.4.1}$$

Then there exists a positive constant C, depending only on L_0, such that for any $z \in L \cap N_\gamma$, $\|z\|^2 \leq C\left(\|z_0\|^2 + \gamma\right)$.

Proof. Let ε be an arbitrarily chosen positive number. Define an operator $F_\varepsilon = F + \varepsilon I$ and a quadratic form

$$\mathcal{F}_\varepsilon(z) = \varepsilon \|z\|^2 + \mathcal{F}(z) = z * F_\varepsilon z .$$

From (2.4.1) we find

$$\varepsilon \|z\|^2 \leq \mathcal{F}_\varepsilon(z) + \gamma. \tag{2.4.2}$$

Then $z = y + z_0$ and

$$\begin{aligned}
\mathcal{F}_\varepsilon(z) &= (y + z_0) F_\varepsilon (y + z_0) \\
&= y * F_\varepsilon y + 2 y * F_\varepsilon z_0 + z_0^* F_\varepsilon z_0.
\end{aligned}$$

Using the inequality $2 y * F_\varepsilon z_0 \leq \varepsilon \|y\|^2 + \varepsilon^{-1} \|F_\varepsilon z_0\|^2$ we obtain

$$\mathcal{F}_\varepsilon(z) \leq \mathcal{F}(y) + 2\varepsilon \|y\|^2 + \varepsilon^{-1} \|F_\varepsilon\|^2 \|z_0\|^2 + \|F_\varepsilon\|^2 \|z_0\|^2 .$$

Take $\varepsilon < \delta/2$. Then

$$\mathcal{F}(y) + 2\varepsilon \|y\|^2 \leq -(\delta - 2\varepsilon)\|y\|^2 \leq 0 .$$

Therefore, $\mathcal{F}_\varepsilon(z) \leq C_0 \|z_0\|^2$ with $C_0 = \varepsilon^{-1} \|F_\varepsilon\|^2 + \|F_\varepsilon\|$. Hence, from (2.4.2) follows the conclusion of the lemma.

Let us now proceed with the proof of Lemma 2.3. We must prove that existence of nonnegative finite measures μ_j, such that $\mu_j(\mathbb{R} \setminus \mathbb{T}) = 0$ and for some constant $\varepsilon > 0$

$$\sum_{j=1}^{N} \int_0^{+\infty} \Pi_j(\omega, \tau) d\mu_j(\tau) + \varepsilon I_m \leq 0 , \tag{2.4.3}$$

where the Hermitian matrix $\Pi(\omega, \tau)$ is defined by the formula

$$\xi^* \Pi_j(\omega, \tau) \xi = \mathcal{F}_j(-W(i\omega)\xi, \xi, -W(i\omega)\xi e^{-i\omega\tau}, \xi e^{-i\omega\tau}) , \tag{2.4.4}$$

implies the validity for any process $z(\cdot) \in L \cap M_\gamma^\infty$ of the estimate

$$\|z(\cdot)\|^2 \leq \lambda \left(\|\alpha(\cdot)\|^2 + \sum_{j=1}^{N} \gamma_j \right) . \tag{2.4.5}$$

In order to apply Lemma 2.11, it is convenient to rewrite each of the Hermitian forms in the right-hand side of (2.4.4) as a Hermitian form $\tilde{\mathcal{F}}_j(\tilde{\sigma}, \tilde{\xi})$ of the variables $\tilde{\sigma} \triangleq -W(i\omega)\tilde{\xi}$ and $\tilde{\xi}$ with the matrix

$$\tilde{F}_j(\omega) = \begin{bmatrix} F_{11j} + F_{33j} + 2F_{13j}\cos\omega\tau & F_{12j} + F_{34j} + F_{14j}e^{-i\omega\tau} + F_{23j}e^{i\omega\tau} \\ F_{12j} + F_{34j} + F_{14j}e^{i\omega\tau} + F_{23j}e^{-i\omega\tau} & F_{22j} + F_{44j} + 2F_{24j}\cos\omega\tau \end{bmatrix}.$$

Here

$$F_j = \begin{bmatrix} F_{11j} & F_{12j} & F_{13j} & F_{14j} \\ F_{12j}^* & F_{22j} & F_{23j} & F_{24j} \\ F_{13j}^* & F_{23j}^* & F_{33j} & F_{34j} \\ F_{14j}^* & F_{24j}^* & F_{34j}^* & F_{44j} \end{bmatrix}$$

is the matrix of the quadratic form $\mathcal{F}_j(\sigma_1,\xi_1,\sigma_2,\xi_2)$. There is no loss of generality in assuming that the matrices F_{13j} and F_{24j} are symmetric.

Recall that stable processes (denoted earlier by Z) in system (2.1.2)-(2.1.3) form a Hilbert space $L^2(0;\infty)$, which will play the role of the space Z in Lemma 2.11. The sets denoted earlier by L and L_0 will play the role of the corresponding subspaces in Lemma 2.11. The quadratic form $\mathcal{F}(z)$ can be taken as

$$\mathcal{F}(z) = \sum_{j=1}^{N} \int_{-\infty}^{+\infty} \tilde{F}_j(\tilde{\sigma}(i\omega),\tilde{\xi}(i\omega))d\omega.$$

The estimate in the conclusion of Lemma 2.3 is the same as in the conclusion of Lemma 2.11. By Lemma 2.11 it holds if

$$\exists \delta > 0: \sum_{j=1}^{N} \int_{-\infty}^{+\infty} \tilde{F}_j(\tilde{\sigma}(i\omega),\tilde{\xi}(i\omega))d\omega \le -\delta\|z\|^2. \qquad (2.4.6)$$

Since $|W(i\omega)|$ is bounded by a constant,

$$\|z\|^2 = \int_{0}^{+\infty} |z(t)|^2 dt = \int_{-\infty}^{+\infty} \left|\begin{bmatrix} \tilde{\sigma}(i\omega) & \tilde{\xi}(i\omega) \end{bmatrix}\right|^2 d\omega.$$

The condition (2.4.6) is equivalent to

$$\exists \delta > 0: \sum_{j=1}^{N} \int_{-\infty}^{+\infty} \tilde{F}_j(\tilde{\sigma}(i\omega),\tilde{\xi}(i\omega))d\omega \le -\delta \int_{-\infty}^{+\infty} |\tilde{\xi}(i\omega)|^2 d\omega. \qquad (2.4.7)$$

Since

$$\mathcal{F}_j(\tilde{\sigma}, \tilde{\xi}) = \begin{bmatrix} \tilde{\sigma} \\ \tilde{\xi} \end{bmatrix}^* \tilde{F}_j(\omega) \begin{bmatrix} \tilde{\sigma} \\ \tilde{\xi} \end{bmatrix},$$

the condition (2.4.7) holds if

$$\exists \delta > 0 : -\sum_{j=1}^{N} \Pi_j(\omega) \geq \delta I_m \qquad (2.4.8)$$

where

$$\Pi_j(\omega) = \begin{bmatrix} -W(i\omega) \\ I_m \end{bmatrix}^* \tilde{F}_j(\omega) \begin{bmatrix} -W(i\omega) \\ I_m \end{bmatrix}.$$

However, this holds if (2.4.3) holds, which completes the required chain of implications and thus the proof of the lemma.

In essence, this lemma establishes that delay-integral-quadratic constraints are a special case of the integral-quadratic constraints in frequency domain.

2.5 Proof of Lemma 2.7

Let M be the essential supremum of the function $|z(t)|$. Define the functionals:

$$\Phi[z(\cdot)] = \sum_{j=1}^{N} \int_0^{+\infty} \int_0^{+\infty} \mathcal{F}_j(z(t), z(t-\tau)) dt d\mu(\tau)$$

$$\Phi_{t_1', t_1}^{t_2', t_2}[z(\cdot)] = \sum_{j=1}^{N} \int_{t_1'}^{t_2'} \int_{t_1}^{t_2} \mathcal{F}_j(z(t), z(t-\tau)) dt d\mu(\tau).$$

Further, define the truncation:

$$\xi_T(t) = \begin{cases} 0 & \text{if } t > T \\ \xi(t) & \text{if } t \leq T. \end{cases}$$

Next, define the truncated processes:

$$z_T = [\sigma_T(\cdot), \xi_T(\cdot)] \in \mathsf{L}_s;$$
$$z_T^0 = [\sigma_T(\cdot) - \alpha(\cdot), \xi_T(\cdot)] \in \mathsf{L}_s^0.$$

Clearly, for $t \le T$, $z_T(t) = z(t)$, and for all t and all nonnegative T, $\left| z_T^0(t) \right| \le M_1$, where M_1 is a constant. For $t \ge T$ we have

$$\sigma_T^0(t) = \int_0^T K(t-s)\xi(s)ds .$$

Therefore, taking into account the condition (2.1.5), we find

$$\left| z_T^0(t) \right| \le C_1 e^{-\beta t} \quad \text{with } C_1 = M + CM/\beta . \tag{2.5.1}$$

Let us denote by F_j the matrix of the form $\mathcal{F}_j$, expand this form, and take advantage of the inequality $2|a*b| \le \delta |a|^2 + \delta^{-1}|b|^2$ for any two vectors a and b. We obtain

$$\mathcal{F}(z_T(t), z(t-\tau)) = \mathcal{F}(z_T^0(t), z_T^0(t-\tau)) + \mathcal{F}(\alpha(t), 0, \alpha(t-\tau), 0)$$
$$+ 2[\alpha*(t) \quad 0 \quad \alpha*(t-\tau) \quad 0] F_j \begin{bmatrix} z_T^0(t) \\ z_T^0(t-\tau) \end{bmatrix}$$
$$\le \delta \left[\left| z_T^0(t) \right|^2 + \left| z_T^0(t-\tau) \right|^2 \right] + \mathcal{F}(z_T^0(t), z_T^0(t-\tau))$$
$$+ c_0 \left[|\alpha(t)|^2 + |\alpha(t-\tau)|^2 \right].$$

Moreover, δ can be chosen to be arbitrarily small.

Therefore, there exist positive constants C_2 and C_3, same for all values of T, such that

$$\Phi_{0,0}^{+\infty, T}[z(\cdot)] = \Phi_{0,0}^{+\infty, T}[z_T(\cdot)] \le \delta C_2 \left\| z_T^0(\cdot) \right\|^2 + \Phi_{0,0}^{+\infty, T}[z_T^0(\cdot)] + C_3 \|\alpha\|^2 .$$

Choose $\delta C_2 < \varepsilon / 4$, where $\varepsilon > 0$ is the same as in the frequency condition (2.2.4). For any process $z(\bullet) \in \mathsf{Z}_s$, there exists a positive constant C_4 such that for all t_1, t_2, and τ

$$\int_{t_1}^{t_2} \mathcal{F}(z(t), z(t-\tau))dt \le C_4 \|z(\cdot)\|^2 .$$

Let T^* be a positive number such that

$$C_4 \sum_{j=1}^{N} \mu_j \left([T^*; +\infty) \right) < \frac{\varepsilon}{4}.$$

We have

$$\Phi_{0,0}^{+\infty,T} \left[z_T^0(\cdot) \right] = -\Phi_{0,T}^{+\infty,+\infty} \left[z_T^0(\cdot) \right] + \Phi \left[z_T^0(\cdot) \right].$$

It follows from the FC (2.2.4) that $\Phi[z_T^0(\cdot)] \leq -\varepsilon \left\| z_T^0(\cdot) \right\|^2$, which implies

$$\left| \Phi_{0,T}^{+\infty,+\infty} \left[z_T^0(\cdot) \right] \right| \leq \left| \Phi_{0,T}^{T^*,+\infty} \left[z_T^0(\cdot) \right] \right| + \left| \Phi_{T^*,T}^{T^*,T+T^*} \left[z_T^0(\cdot) \right] \right|$$

$$\leq \frac{\varepsilon}{4} \left\| z_T^0(\cdot) \right\|^2 + \left| \Phi_{0,T+T^*}^{T^*,+\infty} \left[z_T^0(\cdot) \right] \right| + \left| \Phi_{0,T}^{T^*,T+T^*} \left[z_T^0(\cdot) \right] \right|.$$

For $t \geq T + T^*$ and $0 \leq \tau \leq T^*$, we have $t \geq t - \tau \geq T$. Recalling (2.5.1), we conclude that there exists a positive constant C_5, such that for all T we have

$$\left| \Phi_{0,T+T^*}^{T^*,+\infty} \left[z_T^0(\cdot) \right] \right| \leq C_5.$$

By virtue of the inequality $\left| z_T^0(t) \right| \leq M_1$, we find that there exists a positive constant C_6, such that

$$\left| \Phi_{0,T}^{T^*,T+T^*} \left[z_T^0(\cdot) \right] \right| \leq C_6.$$

Set the variable T to be an arbitrary term t_k of the sequence in (2.1.6). Then from the above inequalities we deduce

$$-\Gamma \leq \Phi_{0,0}^{+\infty,t_k} \left[z(\cdot) \right]$$

$$\leq C_3 \left\| \alpha(\cdot) \right\|^2 + \frac{\varepsilon}{4} \left\| z_{t_k}^0(\cdot) \right\|^2 + \Phi_{0,0}^{+\infty,t_k} \left[z_{t_k}^0(\cdot) \right]$$

$$\leq C_3 \left\| \alpha(\cdot) \right\|^2 + \frac{\varepsilon}{4} \left\| z_{t_k}^0(\cdot) \right\|^2 - \varepsilon \left| z_{t_k}^0(\cdot) \right|^2 + \frac{\varepsilon}{4} \left\| z_{t_k}^0(\cdot) \right\|^2 + C_5 + C_6.$$

Here $\Gamma = \sum_{j=1}^{N} \gamma_j \mu_j \left(\mathbb{R}_+ \right)$.

It follows that

$$\left\| z_{t_k}^0 \right\|^2 \le \frac{2}{\varepsilon}\left(\Gamma + C_3 \left\| \alpha(\cdot) \right\|^2 + C_5 + C_6 \right).$$

Since $z_{t_k}(\cdot) = [\alpha(\cdot), \quad 0] + z_{t_k}^0(\cdot)$, we find that

$$\int_0^{t_k} |z(t)|^2 \, dt \le \left\| z_{t_k} \right\|^2 \le \text{const}.$$

In the limit as $t_k \to +\infty$, we conclude that $|z(\cdot)| \in L^2(0, +\infty)$ QED.

Chapter 3
Stability Multipliers

3.1 General Form for Stability Multipliers for SISO Systems

Consider the feedback system with the linear block given by a Volterra integral equation

$$\sigma(t) = \alpha(t) + K_0\xi(t) + \int_0^t K(t-s)\xi(s)ds .\tag{3.1.1}$$

and the nonlinear block given by

$$\xi(t) = \varphi(\sigma(t),t) .\tag{3.1.2}$$

In these equations $\sigma(t)\in \mathbb{R}$, $\alpha(t)\in \mathbb{R}$, $\xi(t)\in \mathbb{R}$.

It will be assumed that the nonlinearity $\varphi(\sigma,t)$ satisfies the slope restriction condition

$$0 \le \frac{\varphi(\sigma_1,t) - \varphi(\sigma_2,t)}{\sigma_1 - \sigma_2} \le \kappa, \varphi(0,t) \equiv 0 \tag{3.1.3}$$

and that there exists a nonempty set $\mathbb{T}$, such that for all $\tau \in \mathbb{T}$ and for all values of σ and t the following inequality holds:

$$\sigma\left[\varphi(\sigma,t-\tau) - \varphi(\sigma,t)\right] \ge 0 .\tag{3.1.4}$$

D. Altshuller: Frequency Domain Criteria for Absolute Stability, LNCIS 432, pp. 43–80.
springerlink.com   © Springer-Verlag London 2013

Clearly this condition implies

$$\int_0^\sigma \varphi(\sigma, t - \tau)d\sigma \geq \int_0^\sigma \varphi(\sigma, t)d\sigma \,,$$

i.e., that the condition 4) of Lemma 2.10 is satisfied. As mentioned in the discussion following this lemma, this condition is satisfied by functions, periodic in t with period τ (they are considered in Chap. 4) and functions of only one variable discussed later in this chapter. In this section we consider the general case and the dual case with the inequality (3.1.4) reversed.

The results of the chapter will be given in terms of the frequency response of the linear block $W(i\omega)$ defined using the Fourier transform of its kernel:

$$W(i\omega) = -K_0 - \int_0^{+\infty} K(t)e^{-i\omega t}\,dt \,. \tag{3.1.5}$$

The results will be proved using delay-integral-quadratic constraints that will be obtained from Lemma 2.10. To this end, we need to understand the implications of combining the inequalities (3.1.3) and (3.1.4).

Lemma 3.1. *Assume that the function $\varphi(\sigma, t)$ satisfies the conditions (3.1.3) and (3.1.4). Denote $\eta(t) \triangleq \kappa\sigma(t) - \varphi(\sigma(t), t)$. Let $\sigma(t)$ be a measurable function such that $\sigma(t) \equiv 0$ for $t<0$. Then the following inequality holds for every number b:*

$$\int_0^b [\eta(t) - \eta(t - \tau)]\varphi(\sigma(t), t)\,dt \geq 0 \,. \tag{3.1.6}$$

If, in addition, the function $\varphi(\sigma, t)$ is odd in σ, then the following inequality holds for every number b:

$$\int_0^b [\eta(t) + \eta(t - \tau)]\varphi(\sigma(t), t)\,dt \geq 0 \,. \tag{3.1.7}$$

The proof of this lemma is given in Subsect. 3.5.1.

Let us state a lemma, dual to Lemma 3.1, concerning the case with the inequality (3.1.4) reversed.

Lemma 3.2. *Assume that the function* $\varphi(\sigma, t)$ *satisfies the condition (3.1.3) and there exists a nonempty set* $\mathbb{T}$, *such that for all* $\tau \in \mathbb{T}$ *and for all values of* σ *and* t *the following inequality holds:*

$$\sigma[\varphi(\sigma, t - \tau) - \varphi(\sigma, t)] \le 0. \tag{3.1.8}$$

Let $\sigma(t)$ *be a measurable function such that* $\sigma(t) \equiv 0$ *for* $t < 0$. *Denote* $\xi(t) = \varphi(\sigma(t), t)$. *Then the following inequality holds for every number* b:

$$\int_0^b [\xi(t) - \xi(t - \tau)][\kappa \sigma(t) - \xi(t)] dt \ge 0. \tag{3.1.9}$$

If, in addition, the function $\varphi(\sigma, t)$ *is odd in* σ *then the following inequality holds for every number* b:

$$\int_0^b [\xi(t) + \xi(t - \tau)][\kappa \sigma(t) - \xi(t)] dt \ge 0. \tag{3.1.10}$$

The proof of this lemma is given in the Subsect. 3.5.2.
Now we proceed with the formulation and proof of the main results of this section.

Theorem 3.3. *Assume the following:*

1) The linear block (3.1.1) satisfies the regularity conditions;
2) The nonlinearity $\varphi(\sigma, t)$ *satisfies the slope restriction condition (3.1.3);*

3) The set $\mathbb{T}$, *such that for all* $\tau \in \mathbb{T}$ *and for all values of* σ *and* t

$$\sigma[\varphi(\sigma, t - \tau) - \varphi(\sigma, t)] \ge 0, \tag{3.1.11}$$

is nonempty;
4) There exists a nonnegative measure $\mu(\tau)$, *such that* $\mu(\mathbb{R} \setminus \mathbb{T}) = 0$, $\mu(\mathbb{R}_+) < 1$, *and for all real values of* ω

$$\text{Re}\left\{\left[\kappa^{-1} + W(i\omega)\right]\left[1 - \int_0^{+\infty} e^{i\omega\tau} d\mu(\tau)\right]\right\} \ge \varepsilon > 0. \tag{3.1.12}$$

Then for all functions $\sigma(\cdot)$ and $\xi(\cdot)$, satisfying both (3.1.1) and (3.1.2), $\sigma(\cdot) \in L^2(0;+\infty)$ and, furthermore, there exists a positive constant λ, independent of the function $\alpha(\cdot)$, such that $\|\sigma(\cdot)\| \le \lambda \|\alpha(\cdot)\|$.

Proof. The conclusion of the theorem is equivalent to the statement that the system (3.1.1)-(3.1.2) is absolutely stable. The proof consists of the following steps. First, using Lemma 3.1, we establish that the nonlinearity satisfies two delay-integral-quadratic constraints. Next, we perform the computation described in Chapter 2 to derive the frequency condition. The final step is the verification of the minimal stability.

Let us proceed with the first step. Define two quadratic forms:

$$\mathcal{F}_1\left(\sigma_1, \xi_1\right) = \xi_1\left(\sigma_1 - \kappa^{-1}\xi_1\right);$$
$$\mathcal{F}_2\left(\sigma_1, \xi_1, \sigma_2, \xi_2\right) = \left(\kappa\sigma_1 - \xi_1 - \kappa\sigma_2 + \xi_2\right)\xi_1.$$

The slope restriction condition implies that

$$\mathcal{F}_1(\sigma(t), \xi(t)) \ge 0. \tag{3.1.13}$$

Lemma 3.1 implies that the following inequality holds for an arbitrary t_k and for all $\tau \in \mathcal{T}$:

$$\int_0^{t_k} \mathcal{F}_2\left(\sigma(t), \xi(t), \sigma(t-\tau), \xi(t-\tau)\right) dt \ge 0. \tag{3.1.14}$$

Computation of the matrices $\Pi_j(\omega, \tau)$ as described in Chap. 2 yields

$$\Pi_1(\omega, \tau) = -\text{Re}\left[\kappa^{-1} + W(i\omega)\right];$$
$$\Pi_2(\omega, \tau) = -\text{Re}\left\{\left[\kappa^{-1} + W(i\omega)\right]\left(1 - e^{i\omega\tau}\right)\right\}.$$

The frequency condition now takes the following form. There exist nonnegative measures $\mu_1(\tau)$ and $\mu_2(\tau)$, such that for all real values of ω

$$\text{Re}\left\{\left[\kappa^{-1} + W(i\omega)\right]\left[\sum_{j=1}^2 \mu_j\left(\mathbb{R}_+\right) - \int_0^{+\infty} e^{i\omega\tau} d\mu_2(\tau)\right]\right\} \ge \varepsilon > 0.$$

We can set $\mu_1\left(\mathbb{R}_+\right) + \mu_2\left(\mathbb{R}_+\right) = 1$ (without loss of generality), $\mu(A) = \mu_2(A)$ for any set $A \subset \mathbb{R}$ and obtain the frequency condition in the assumption 4) of the theorem.

Verification of minimal stability is given in Subsect. 3.5.3. By Theorem 2.6 the system (3.1.1)-(3.1.2) is absolutely stable, which is equivalent to the conclusion of the theorem. The proof is complete.

If an additional assumption that the nonlinearity is odd in σ is introduced, it is possible to relax the requirement that the measure $\mu(\tau)$ is nonnegative.

Theorem 3.4. *Assume that hypotheses 1), 2), and 3) of Theorem 3.3 are satisfied and, in addition, the function $\varphi(\sigma,t)$ is odd in σ. Assume further that there exists a signed measure $\mu(\tau)$, such that $\mu(\mathbb{R}\setminus\mathbb{T})=0$, $\mu(\mathbb{R}_+)<1$ and for all real values of ω*

$$\mathrm{Re}\left\{\left[\kappa^{-1}+W(i\omega)\right]\left[1-\int_0^{+\infty}e^{i\omega\tau}d\mu(\tau)\right]\right\}\geq\varepsilon>0. \tag{3.1.15}$$

Then for all functions $\sigma(\cdot)$ and $\xi(\cdot)$, satisfying both (3.1.1) and (3.1.2), $\sigma(\cdot)\in L^2(0;+\infty)$ and, furthermore, there exists a positive constant λ, independent of the function $\alpha(\cdot)$, such that $\|\sigma(\cdot)\|\leq\lambda\|\alpha(\cdot)\|$.

Proof. The proof proceeds along the same steps as Theorem 3.3. Define the same quadratic forms $\mathcal{F}_1(\sigma_1,\xi_1)$ and $\mathcal{F}_2(\sigma_1,\xi_1,\sigma_2,\xi_2)$. The inequalities (3.1.13) and (3.1.14) hold.

Define the additional quadratic form

$$\mathcal{F}_3\left(\sigma_1,\xi_1,\sigma_2,\xi_2\right)=\left(\kappa\sigma_1-\xi_1+\kappa\sigma_2-\xi_2\right)\xi_1.$$

Since the function $\varphi(\sigma,t)$ is odd in σ, Lemma 3.1 implies that the following inequality holds for an arbitrary t_k and for all $\tau\in\mathbb{T}$:

$$\int_0^{t_k}\mathcal{F}_3\left(\sigma(t),\xi(t),\sigma(t-\tau),\xi(t-\tau)\right)dt\geq0. \tag{3.1.16}$$

Computation of the matrix $\Pi_3(\omega,\tau)$ yields

$$\Pi_3(\omega,\tau)=-\mathrm{Re}\left\{\left[\kappa^{-1}+W(i\omega)\right]\left(1+e^{i\omega\tau}\right)\right\}.$$

Because of this additional constraint, the frequency condition takes the following form. There exist nonnegative measures $\mu_1(\tau)$, $\mu_2(\tau)$, and $\mu_3(\tau)$, such that for all real values of ω

$$\text{Re}\left\{\left[\kappa^{-1}+W(i\omega)\right]\left[\sum_{j=1}^{3}\mu_j\left(\mathbb{R}_+\right)-\int_0^{+\infty}e^{i\omega\tau}d\mu_2(\tau)+\int_0^{+\infty}e^{i\omega\tau}d\mu_3(\tau)\right]\right\}\geq\varepsilon>0.$$

Once again, set without loss of generality

$$\sum_{j=1}^{3}\mu_j\left(\mathbb{R}_+\right)=1$$

and define $\mu(A)=\mu_2(A)-\mu_3(A)$ for any set $A\subset\mathbb{R}_+$. This yields the frequency condition of the theorem. The proof concludes in the same way as the proof of Theorem 3.3.

Note that if the function $\varphi(\sigma,t)$ is nonincreasing in t for positive σ and nondecreasing in t for negative σ, then $\mathbb{R}\setminus\mathbb{T}=\varnothing$.

Let us now proceed with formulation and proof of the results for the case when the inequality (3.1.4) is reversed.

Theorem 3.5. *Assume the following:*

1) The linear block (3.1.1) satisfies the regularity conditions;
2) The nonlinearity $\varphi(\sigma,t)$ satisfies the slope restriction condition (3.1.3);

3) The set $\mathbb{T}$, such that for all $\tau\in\mathbb{T}$ and for all values of σ and t

$$\sigma\left[\varphi(\sigma,t-\tau)-\varphi(\sigma,t)\right]\leq0,\tag{3.1.17}$$

is nonempty;
4) There exists a nonnegative measure $\mu(\tau)$, such that $\mu\left(\mathbb{R}\setminus\mathbb{T}\right)=0$, $\mu(\mathbb{R}_+)<1$ and for all real values of ω

$$\text{Re}\left\{\left[\kappa^{-1}+W(i\omega)\right]\left[1-\int_0^{+\infty}e^{-i\omega\tau}d\mu(\tau)\right]\right\}\geq\varepsilon>0.\tag{3.1.18}$$

Then for all functions $\sigma(\cdot)$ and $\xi(\cdot)$, satisfying both (3.1.1) and (3.1.2), $\sigma(\cdot)\in L^2(0;+\infty)$ and, furthermore, there exists a positive constant λ, independent of the function $\alpha(\cdot)$, such that $\|\sigma(\cdot)\|\leq\lambda\|\alpha(\cdot)\|$.

Proof The proof of this theorem proceeds along the same steps as Theorem 3.3. First, we define two quadratic forms:

$$\mathcal{F}_1\left(\sigma_1,\xi_1\right)=\xi_1\left(\sigma_1-\kappa^{-1}\xi_1\right);$$
$$\mathcal{F}_2\left(\sigma_1,\xi_1,\sigma_2,\xi_2\right)=\left(\kappa\sigma_1-\xi_1\right)\left(\xi_1-\xi_2\right).$$

The slope restriction condition implies that

$$\mathcal{F}_1(\sigma(t), \xi(t)) \geq 0 . \tag{3.1.19}$$

Lemma 3.2 implies that the following inequality holds for an arbitrary t_k and for all $\tau \in \mathbb{T}$:

$$\int_0^{t_k} \mathcal{F}_2\big(\sigma(t), \xi(t), \sigma(t-\tau), \xi(t-\tau)\big) dt \geq 0 . \tag{3.1.20}$$

Computation of the matrices $\Pi_j(\omega, \tau)$ as described in Chap. 2 yields

$$\Pi_1(\omega, \tau) = -\mathrm{Re}\big[\kappa^{-1} + W(i\omega)\big];$$
$$\Pi_2(i\omega, \tau) = -\mathrm{Re}\big\{\big[\kappa^{-1} + W(i\omega)\big]\big(1 - e^{-i\omega\tau}\big)\big\}.$$

The frequency condition now takes the following form. There exist nonnegative measures $\mu_1(\tau)$ and $\mu_2(\tau)$, such that for all real values of ω

$$\mathrm{Re}\left\{\big[\kappa^{-1} + W(i\omega)\big]\left[\sum_{j=1}^2 \mu_j(\mathbb{R}_+) - \int_0^{+\infty} e^{-i\omega\tau} d\mu_2(\tau)\right]\right\} \geq \varepsilon > 0 .$$

The proof now concludes in the same way as the proof of Theorem 3.3.

Just as in the case of Theorem 3.3, if the nonlinearity is odd in σ, it is possible to relax the assumption that the measure $\mu(\tau)$ is nonnegative. This yields the result dual to Theorem 3.4.

Theorem 3.6. *Assume that hypotheses 1), 2), and 3) of Theorem 3.5 are satisfied and, in addition, the function $\varphi(\sigma, t)$ is odd in σ. Assume further that there exists a signed measure $\mu(\tau)$, such that $\mu(\mathbb{R} \setminus \mathbb{T}) = 0$, $\mu(\mathbb{R}_+) < 1$ and for all real values of ω*

$$\mathrm{Re}\left\{\big[\kappa^{-1} + W(i\omega)\big]\left[1 - \int_0^{+\infty} e^{-i\omega\tau} d\mu(\tau)\right]\right\} \geq \varepsilon > 0. \tag{3.1.21}$$

Then for all functions $\sigma(\cdot)$ and $\xi(\cdot)$, satisfying both (3.1.1) and (3.1.2), $\sigma(\cdot) \in L^2(0; +\infty)$ and, furthermore, there exists a positive constant λ, independent of the function $\alpha(\cdot)$, such that $\|\sigma(\cdot)\|^2 \leq \lambda\|\alpha(\cdot)\|$.

Proof. This theorem is proved by the same arguments as Theorem 3.4, but with the help of Lemma 3.2. As a first step, we define the same quadratic forms $\mathcal{F}_1(\sigma_1,\xi_1)$ and $\mathcal{F}_2(\sigma_1,\xi_1,\sigma_2,\xi_2)$ as in the proof of Theorem 3.5. The inequalities (3.1.19) and (3.1.20) hold.

In addition, we define the following quadratic form:

$$\mathcal{F}_3(\sigma_1,\xi_1,\sigma_2,\xi_2)=(\kappa\sigma_1-\xi_1)(\xi_1+\xi_2).$$

Lemma 3.2 implies that the following inequality holds for an arbitrary t_k and for all $\tau\in\mathbb{T}$:

$$\int_0^{t_k}\mathcal{F}_3\big(\sigma(t),\xi(t),\sigma(t-\tau),\xi(t-\tau)\big)\,dt\geq0.\tag{3.1.22}$$

Computation of the matrix $\Pi_3(\omega,\tau)$ yields

$$\Pi_3(\omega,\tau)=-\mathrm{Re}\left\{\left[\kappa^{-1}+W(i\omega)\right]\left(1+e^{-i\omega\tau}\right)\right\}.$$

The frequency condition now takes the form: There exist nonnegative measures $\mu_1(\tau),\mu_2(\tau)$, and $\mu_3(\tau)$, such that for all real values of ω

$$\mathrm{Re}\left\{\left[\kappa^{-1}+W(i\omega)\right]\left[\sum_{j=1}^{3}\mu_j(\mathbb{R}_+)-\int_0^{+\infty}e^{-i\omega\tau}d\mu_2(\tau)+\int_0^{+\infty}e^{-i\omega\tau}d\mu_3(\tau)\right]\right\}\geq\varepsilon>0.$$

The proof concludes in exactly the same way as the proof of Theorem 3.4.

Note that if the function $\varphi(\sigma,t)$ is nondecreasing in t for positive σ and nonincreasing in t for negative σ, then $\mathbb{R}\setminus\mathbb{T}=\varnothing$.

3.2 Multipliers for Stationary SISO Systems

We now turn our attention to stationary systems, i.e., the systems with nonlinearity depending only on the variable σ and satisfying the slope restriction condition

$$0\leq\frac{\varphi(\sigma_1)-\varphi(\sigma_2)}{\sigma_1-\sigma_2}\leq\kappa,\varphi(0)=0.\tag{3.2.1}$$

These nonlinearities satisfy the hypotheses of both Lemma 3.1 and Lemma 3.2. For the sake of making references easy, let us state a separate lemma.

Lemma 3.7. *Assume that the function* $\varphi(\sigma)$ *satisfies the slope restriction condition* (3.2.1). *Denote* $\eta(t) \triangleq \kappa\sigma(t) - \varphi(\sigma(t))$ *and* $\xi(t) = \varphi(\sigma(t))$. *Then for any measurable function* $\sigma(t)$, *the following inequalities hold for every real number* b:

$$\int_0^b [\eta(t) - \eta(t-\tau)] \varphi(\sigma(t)) \, dt \geq 0;$$

$$\int_0^b [\xi(t) - \xi(t-\tau)] \eta(t) dt \geq 0.$$

If, in addition, the function $\varphi(\sigma)$ *is odd then the following inequalities holds for every real number* b:

$$\int_0^b [\eta(t) + \eta(t-\tau)] \varphi(\sigma(t)) \, dt \geq 0;$$

$$\int_0^b [\xi(t) + \xi(t-\tau)] \eta(t) dt \geq 0.$$

Proof. Note that the functions $\varphi(\sigma)$ and $\kappa\sigma - \varphi(\sigma)$ satisfy automatically the conditions 1)-4) of Lemma 2.10. Since they are nondecreasing, they satisfy the condition 5) of Lemma 2.10 with $H(u,v) \equiv 0$. Application of Lemma 2.10 yields the desired conclusions.

In this section we shall first prove the classical result of Zames and Falb [173] using the method of delay-integral-quadratic constraints. Then, this result will be strengthened for the case when the kernel of the linear block is an absolutely continuous function, which will lead to the addition of the Popov term to the frequency condition. Finally, we will prove a result for the case of differentiable nonlinearities. This result combines the Yakubovich criterion with the result of Zames and Falb. It will be interesting to observe how introduction of additional assumptions about the nonlinearity yields new constraints, thus strengthening the frequency condition – the usual tradeoff in the theory of absolute stability.

3.2.1 Zames-Falb Multipliers

Let us proceed with the formulation and proof of the main result. It was first proved by Zames and Falb [173] using a different method.

Theorem 3.8. *Assume the following:*

1) The linear block (3.1.1) *satisfies the regularity conditions;*
2) The function $\varphi(\sigma)$ *satisfies the slope restriction condition* (3.1.3);

3) There exists a nonnegative measure $\mu(\tau)$, such that $\mu(\mathbb{R}) < 1$ and for all real values of ω

$$\text{Re}\left\{\left[\kappa^{-1} + W(i\omega)\right]\left[1 - \int_{-\infty}^{+\infty} e^{i\omega\tau} d\mu(\tau)\right]\right\} \geq \varepsilon > 0. \tag{3.2.2}$$

Then for all functions $\sigma(\cdot)$ and $\xi(\cdot)$, satisfying both (3.1.1) and (3.1.2), $\sigma(\cdot) \in L^2(0;+\infty)$ and, furthermore, there exists a positive constant λ, independent of the function $\alpha(\cdot)$, such that $\|\sigma(\cdot)\| \leq \lambda\|\alpha(\cdot)\|$.

Proof. Define three quadratic forms:

$$\mathcal{F}_1(\sigma_1,\xi_1) = \xi_1\left(\sigma_1 - \kappa^{-1}\xi_1\right);$$
$$\mathcal{F}_2(\sigma_1,\xi_1,\sigma_2,\xi_2) = \left(\kappa\sigma_1 - \xi_1 - \kappa\sigma_2 + \xi_2\right)\xi_1;$$
$$\mathcal{F}_3(\sigma_1,\xi_1,\sigma_2,\xi_2) = \left(\kappa\sigma_1 - \xi_1\right)\left(\xi_1 - \xi_2\right).$$

The slope restriction condition implies that

$$\mathcal{F}_1(\sigma(t),\xi(t)) \geq 0. \tag{3.2.3}$$

Lemma 3.7 yields the following two inequalities, which hold for all values of t_k and all nonnegative real values of τ:

$$\int_0^{t_k} \mathcal{F}_j(\sigma(t),\xi(t),\sigma(t-\tau),\xi(t-\tau))\,dt \geq 0,\ j = 2,3. \tag{3.2.4}$$

Compute the matrices $\Pi_j(\omega,\tau)$ as described in Chap. 2 to obtain

$$\Pi_1(\omega,\tau) = -\text{Re}\left[\kappa^{-1} + W(i\omega)\right];$$
$$\Pi_2(\omega,\tau) = -\text{Re}\left\{\left[\kappa^{-1} + W(i\omega)\right]\left(1 - e^{i\omega\tau}\right)\right\};$$
$$\Pi_3(\omega,\tau) = -\text{Re}\left\{\left[\kappa^{-1} + W(i\omega)\right]\left(1 - e^{-i\omega\tau}\right)\right\}.$$

The frequency condition now takes the following form. There exist nonnegative measures $\mu_1(\tau)$, $\mu_2(\tau)$, and $\mu_3(\tau)$, such that for all real values of ω

$$\text{Re}\left\{\left[\kappa^{-1} + W(i\omega)\right]\left[\sum_{j=1}^{3}\mu_j(\mathbb{R}_+) - \int_0^{+\infty} e^{i\omega\tau} d\mu_2(\tau) - \int_0^{+\infty} e^{-i\omega\tau} d\mu_3(\tau)\right]\right\} \geq \varepsilon > 0.$$

As before, we can set without loss of generality

$$\sum_{j=1}^{3} \mu_j(\mathbb{R}_+) = 1.$$

Define for any set $A \subset \mathbb{R}$ the measure $\mu(A) = \mu_2(A \cap \mathbb{R}_+) + \mu_3(-A \cap \mathbb{R}_+)$. Then

$$\int_0^{+\infty} e^{i\omega\tau} d\mu_2(\tau) + \int_0^{+\infty} e^{-i\omega\tau} d\mu_3(\tau)$$

$$= \int_0^{+\infty} e^{i\omega\tau} d\mu_2(\tau) - \int_{-\infty}^0 e^{-i\omega\tau} d\mu_3(\tau)$$

$$= \int_{-\infty}^{+\infty} e^{i\omega\tau} d\mu(\tau).$$

Therefore, the frequency condition reduces to the one stated in the hypothesis 3) of the theorem. The proof is completed by verification of minimal stability as described in Subsect. 3.5.3.

Let the measure $\mu(\tau)$ be generated by a nondecreasing function $\vartheta(\tau)$. Then we can rewrite the integral in (3.2.2) in the form of Lebesgue-Stieltjes and apply the canonical decomposition to obtain

$$\int_{-\infty}^{+\infty} e^{i\omega\tau} d\mu(\tau) = \int_{-\infty}^{+\infty} e^{i\omega\tau} d\vartheta(\tau) = \int_{-\infty}^{+\infty} e^{i\omega t} z(t) dt + \sum_{j=1}^{\infty} z_j e^{i\omega b_j},$$

where $z(t)$ is the absolutely continuous component of the function $\vartheta(\tau)$, which is also assumed to have jumps of magnitude z_j at points $\tau = b_j$. This reduces the frequency condition to the one obtained by Zames and Falb [172, 173]. Furthermore, the requirement $\mu(\mathbb{R}) < 1$ is equivalent to the inequality

$$\int_{-\infty}^{+\infty} |z(t)| dt + \sum_{j=1}^{\infty} |z_j| < 1.$$

If the nonlinearity is odd, it is possible, as before, to relax the condition that the measure $\mu(\tau)$ must be nonnegative.

Theorem 3.9. *Assume that the conditions 1) and 2) of Theorem 3.8 are satisfied and the function $\varphi(\sigma)$ is odd. Assume further that there exists a signed measure $\mu(\tau)$, such that $\mu(\mathbb{R}) < 1$ and for all real values of ω*

$$\text{Re}\left\{\left[\kappa^{-1}+W(i\omega)\right]\left[1-\int_{-\infty}^{+\infty}e^{i\omega\tau}d\mu(\tau)\right]\right\}\geq\varepsilon>0. \tag{3.2.5}$$

Then for all functions $\sigma(\cdot)\,and\,\xi(\cdot)$, *satisfying both* (3.1.1) *and* (3.1.2), $\sigma(\cdot)\in L^2(0;+\infty)$ *and, furthermore, there exists a positive constant* λ, *independent of the function* $\alpha(\cdot)$, *such that* $\|\sigma(\cdot)\|\leq\lambda\|\alpha(\cdot)\|$.

Proof. Define the same quadratic forms $\mathcal{F}_1(\sigma_1,\xi_1),\mathcal{F}_2(\sigma_1,\xi_1,\sigma_2,\xi_2)$, and $\mathcal{F}_3(\sigma_1,\xi_1,\sigma_2,\xi_2)$ as in the proof of Theorem 3.8. The inequalities (3.2.3) and (3.2.4) hold.

Define two more quadratic forms:

$$\mathcal{F}_4(\sigma_1,\xi_1,\sigma_2,\xi_2)=(\kappa\sigma_1-\xi_1+\kappa\sigma_2-\xi_2)\xi_1;$$
$$\mathcal{F}_5(\sigma_1,\xi_1,\sigma_2,\xi_2)=(\kappa\sigma_1-\xi_1)(\xi_1+\xi_2).$$

Lemma 3.7 implies that (3.2.4) holds for $j=4,5$.

Compute the matrices $\Pi_j(\omega,\tau)$ to obtain

$$\Pi_4(\omega,\tau)=-\text{Re}\left\{\left[\kappa^{-1}+W(i\omega)\right]\left(1+e^{i\omega\tau}\right)\right\};$$
$$\Pi_5(\omega,\tau)=-\text{Re}\left\{\left[\kappa^{-1}+W(i\omega)\right]\left(1+e^{-i\omega\tau}\right)\right\}.$$

Now the frequency condition takes the following form. There exist nonnegative measures $\mu_j(\tau),1\leq j\leq5$, such that for all real values of ω

$$\text{Re}\left\{\left[\kappa^{-1}+W(i\omega)\right]Z(i\omega)\right\}\geq\varepsilon>0.$$

The multiplier $Z(i\omega)$ is given by

$$Z(i\omega)\triangleq\sum_{j=1}^{5}\mu_j(\mathbb{R}_+)-\int_0^{+\infty}e^{i\omega\tau}d\mu_2(\tau)-\int_0^{+\infty}e^{-i\omega\tau}d\mu_3(\tau)$$
$$+\int_0^{+\infty}e^{i\omega\tau}d\mu_4(\tau)+\int_0^{+\infty}e^{-i\omega\tau}d\mu_5(\tau).$$

Again, without loss of generality we can set

$$\sum_{j=1}^{5}\mu_j(\mathbb{R}_+)=1.$$

Now define for any set $A \subset \mathbb{R}$ two signed measures $\bar{\mu}_2(A) = \mu_2(A) - \mu_4(A)$ and $\bar{\mu}_3(A) = \mu_3(A) - \mu_5(A)$ and then the measure

$$\mu(A) = \bar{\mu}_2(A \cap \mathbb{R}_+) + \bar{\mu}_3(-A \cap \mathbb{R}_+).$$

The proof is concluded in the same way as the proof of Theorem 3.8.

The same argument as the one following the proof of Theorem 3.8 can be applied here as well except that we do not have to impose the requirement for the function $\vartheta(\tau)$ to be nondecreasing – it only has to be of bounded variation and not have a singular component. Therefore, the method of delay-integral-quadratic constraints makes it possible to give a new, simpler proof of the classic result of Zames and Falb.

3.2.2 Case of an Absolutely Continuous Kernel: Popov Criterion

The objective of this subsection is to prove the Popov criterion using the constraints method. It will then be combined with the results of the previous subsection to obtain another result, also originally due to Zames and Falb [172].

Instead of slope restriction, a weaker assumption called the sector condition will be made about the nonlinearity:

$$0 \le \frac{\varphi(\sigma)}{\sigma} \le \kappa, \varphi(0) = 0. \tag{3.2.6}$$

Theorem 3.10 (Popov criterion). *Assume the following:*

1) The linear block (3.1.1) satisfies the regularity conditions;
2) The functions $\alpha(t)$ and $K(t)$ are absolutely continuous, $K_0 = 0$, and $|\dot{\alpha}(\bullet)| \in L^2(0; \infty)$;
3) The function $\varphi(\sigma)$ satisfies the sector condition (3.2.6);
4) There exists a constant θ, such that for all real values of ω

$$\operatorname{Re}\left\{ \left[\kappa^{-1} + W(i\omega) \right] (1 + i\omega\theta) \right\} \ge \varepsilon > 0.$$

Then for all functions $\sigma(\bullet)$ and $\xi(\bullet)$, satisfying both (3.1.1) and (3.1.2), $\sigma(\bullet) \in L^2(0; +\infty)$ and, furthermore, there exists a positive constant λ, independent of the function $\alpha(\bullet)$, such that

$$\left\| \sigma(\bullet) \right\|^2 \le \lambda \left[\left\| \alpha(\bullet) \right\|^2 + 2 \int_0^{\sigma(0)} \varphi(\sigma) d\sigma + \zeta \frac{\sigma^2(0)}{2} \right]. \tag{3.2.7}$$

Proof. In order to prove this theorem, we first have to add the component $\dot{\sigma}$ to the vector $\begin{bmatrix} \sigma & \xi \end{bmatrix}$. Taking advantage of the absolute continuity of the functions $\alpha(t)$ and $K(t)$, we have the equation

$$\dot{\sigma}(t) = \dot{\alpha}(t) + \int_0^t \frac{\partial K(t-s)}{\partial t} \xi(t)ds + K(0)\xi(t).$$

We now proceed with the usual steps of deriving the constraints. Note that σ_1 is now a vector $\begin{bmatrix} \sigma & \dot{\sigma} \end{bmatrix}$. The transfer function is

$$\tilde{\sigma}_1 = \begin{bmatrix} \tilde{\sigma} \\ \tilde{\dot{\sigma}} \end{bmatrix} = \begin{bmatrix} -W(i\omega) \\ -i\omega W(i\omega) \end{bmatrix} \tilde{\xi}.$$

First, as before, define the quadratic form

$$\mathcal{F}_1(\sigma_1, \xi_1) = \xi_1\left(\sigma_1 - \kappa^{-1}\xi_1\right).$$

Sector condition (3.2.6) implies that

$$\mathcal{F}_1(\sigma_1(t), \xi(t)) \geq 0. \tag{3.2.8}$$

Next two constraints will involve the component $\dot{\sigma}$. Define

$$\mathcal{F}_2(\sigma_1, \xi_1) = \dot{\sigma}\xi_1.$$

Sector condition (3.2.6) implies that for arbitrary t_k

$$\int_0^{\sigma(t_k)} \varphi(\sigma)d\sigma \geq 0.$$

Using the substitution $\sigma = \sigma(t)$, we can rewrite the integral on the left-hand side as follows:

$$\int_0^{\sigma(t_k)} \varphi(\sigma)d\sigma = \int_0^{t_k} \varphi(\sigma(t))\frac{d\sigma}{dt}dt + \int_0^{\sigma(0)} \varphi(\sigma)d\sigma = \int_0^{t_k} \xi(t)\dot{\sigma}(t)dt + \int_0^{\sigma(0)} \varphi(\sigma)d\sigma.$$

Denote

$$\gamma_2 = \int_0^{\sigma(0)} \varphi(\sigma)d\sigma.$$

The above considerations imply

$$\int_0^{t_k} \mathcal{F}_2(\sigma_1(t), \xi(t))dt \geq -\gamma_2. \tag{3.2.9}$$

Furthermore, define the quadratic form

$$\mathcal{F}_3(\sigma_1, \xi_1) = (\kappa\sigma - \xi)\dot{\sigma}.$$

Considerations, similar to the above, yield the inequality

$$\int_0^{t_k} \mathcal{F}_3(\sigma_1(t), \xi_1(t))dt \geq -\gamma_3, \tag{3.2.10}$$

where

$$\gamma_3 = \int_0^{\sigma(0)} [\kappa\sigma - \varphi(\sigma)]d\sigma = \kappa\frac{\sigma^2(0)}{2} - \int_0^{\sigma(0)} \varphi(\sigma)d\sigma.$$

The matrices $\Pi_j(\omega, \tau)$ are

$$\Pi_1(\omega, \tau) = -\text{Re}\left[\kappa^{-1} + W(i\omega)\right];$$
$$\Pi_2(\omega, \tau) = -\text{Re}\left\{i\omega\left[\kappa^{-1} + W(i\omega)\right]\right\};$$
$$\Pi_3(\omega, \tau) = \text{Re}\left\{i\omega\left[\kappa^{-1} + W(i\omega)\right]\right\}.$$

The frequency condition now takes the following form. There exist nonnegative measures $\mu_1(\tau)$, $\mu_2(\tau)$, and $\mu_3(\tau)$, such that for all real values of ω

$$\text{Re}\left\{\left[\sum_{j=1}^{3}\mu_j(\tau) - i\omega\int_0^{+\infty}d\mu_2(\tau) + i\omega\int_0^{+\infty}d\mu_3(\tau)\right]\left[\kappa^{-1} + W(i\omega)\right]\right\} \geq \varepsilon > 0.$$

Set

$$\sum_{j=1}^{3}\mu_j(\tau) = 1, \theta = \int_0^{+\infty}d\mu_3(\tau) - \int_0^{+\infty}d\mu_2(\tau)$$

to obtain the frequency condition of the theorem.

Verification of minimal stability is given in the Subsect. 3.5.3.

By Lemma 2.6 the system is absolutely stable, which by definition means validity of the estimate (3.2.7).

This method of proving the Popov criterion was given by Yakubovich in several papers [153, 160, 164, 165].

If the nonlinearity satisfies the slope restriction condition, then the Popov criterion can be combined with the results of the previous Subsection to obtain the following two theorems.

Theorem 3.11. *Assume the following*:

1) The linear block (3.1.1) satisfies the regularity conditions;
2) The functions $\alpha(t)$ and $K(t)$ are absolutely continuous, $K_0 = 0$, and $|\dot{\alpha}(\bullet)| \in L^2(0;\infty)$;

3) The function $\varphi(\sigma)$ satisfies the slope restriction condition

$$0 \le \frac{\varphi(\sigma_1) - \varphi(\sigma_2)}{\sigma_1 - \sigma_2} \le \kappa, \varphi(0) = 0 \; ;$$

4) There exist a constant θ and a nonnegative measure $\mu(\tau)$, such that $\mu(\mathbb{R}) < 1$ and for all real values of ω

$$\mathrm{Re}\left\{ \left[\kappa^{-1} + W(i\omega) \right] \left[1 + i\omega\theta - \int_{-\infty}^{+\infty} e^{i\omega\tau} d\mu(\tau) \right] \right\} \ge \varepsilon > 0 \; . \qquad (3.2.11)$$

Then for all functions $\sigma(\bullet)$ and $\xi(\bullet)$, satisfying both (3.1.1) and (3.1.2), $\sigma(\bullet) \in L^2(0;+\infty)$ and, furthermore, there exists a positive constant λ, independent of the function $\alpha(\bullet)$, such that

$$\|\sigma(\bullet)\|^2 \le \lambda \left[\|\alpha(\bullet)\|^2 + 2 \int_0^{\sigma(0)} \varphi(\sigma)d\sigma + \zeta \frac{\sigma^2(0)}{2} \right]. \qquad (3.2.12)$$

Theorem 3.12. *Assume that the hypotheses 1), 2), and 3) of the theorem 3.11 are satisfied and the function $\varphi(\sigma)$ is odd. Suppose further that there exist a constant θ and a signed measure $\mu(\tau)$, such that $\mu(\mathbb{R}) < 1$ and for all real values of ω the inequality (3.2.11) holds. Then the conclusion of the Theorem 3.11 is true.*

Both of these theorems are proved by combining the arguments of the previous Subsection with the proof of Theorem 3.10. They were first established (using a different method) by Zames and Falb [172].

We can also apply the arguments of the previous subsection and replace the integral in (3.2.11) with the Lebesgue-Stieltjes integral. Assuming further that the function $\vartheta(\tau)$ is absolutely continuous, we obtain the frequency condition in the form proved by Barabanov [19]:

$$\text{Re}\left\{\left[\kappa^{-1}+W(i\omega)\right]\left[1+i\omega\theta-\int_{-\infty}^{+\infty}\vartheta'(\tau)e^{i\omega\tau}d\tau\right]\right\} \geq \varepsilon > 0. \tag{3.2.13}$$

3.2.3 Case of a Differentiable Nonlinearity: Yakubovich Criterion

In this subsection we return to systems with slope restricted nonlinearities, i.e., it will be assumed that the function $\varphi(\sigma)$ satisfies the condition

$$0 \leq \frac{\Delta\varphi(\sigma)}{\Delta\sigma} \leq \kappa, \varphi(0) = 0. \tag{3.2.14}$$

Furthermore, it will be assumed, as in the previous subsection, that the kernel of the linear block is absolutely continuous and, in addition, the function $\varphi(\sigma)$ is differentiable. For systems satisfying these conditions we shall prove the result known as the Yakubovich criterion [148].

Theorem 3.13 (Yakubovich criterion). *Assume the following:*

1) The linear block (3.1.1) satisfies the regularity conditions;
2) The functions $\alpha(t)$ and $K(t)$ are absolutely continuous, $K_0=0$, and
$|\dot{\alpha}(\cdot)| \in L^2(0;\infty)$;

3) The function $\varphi(\sigma)$ is differentiable and satisfies the slope restriction condition (3.2.14);
4) There exist a constant θ_1 and a positive constant θ_2, such that for all real values of ω

$$\text{Re}\left\{\left[\kappa^{-1}+W(i\omega)\right]\left(1+i\omega\theta_1+\theta_2\omega^2\right)\right\} \geq \varepsilon > 0.$$

Then $\sigma(\cdot) \in L^2(0;+\infty)$ for all functions $\sigma(\cdot)$ and $\xi(\cdot)$, satisfying both and (3.1.2), and, furthermore, there exists a positive constant λ, independent of the function $\alpha(\cdot)$, such that

$$\|\sigma(\cdot)\|^2 \leq \lambda\left[\|\alpha(\cdot)\|^2 + 2\int_0^{\sigma(0)}\varphi(\sigma)d\sigma + \zeta\frac{\sigma^2(0)}{2}\right]. \tag{3.2.15}$$

Proof. First, note that conditions 2) and 3) imply that conditions 2) and 3) of Theorem 3.10 hold. Therefore, we can use the arguments from the proof of that theorem without any changes.

Next, add a component $\dot{\xi}$ to the vector $\begin{bmatrix} \sigma & \dot{\sigma} & \xi \end{bmatrix}$, which we obtained in the previous Subsection by adding the component $\dot{\sigma}$ to the vector $\begin{bmatrix} \sigma & \xi \end{bmatrix}$. Since the function $\varphi(\sigma)$ is differentiable, we have $\dot{\xi}(t) = \varphi'(\sigma)\dot{\sigma}(t)$. In the limit as $\Delta\sigma \to 0$, the slope restriction condition (3.2.14) becomes

$$0 \le \frac{\dot{\xi}}{\dot{\sigma}} \le \kappa < \infty. \tag{3.2.16}$$

Define, in addition to the quadratic forms $\mathcal{F}_1(\sigma_1,\xi_1)$, $\mathcal{F}_2(\sigma_1,\xi_1)$, and $\mathcal{F}_3(\sigma_1,\xi_1)$ from the previous subsection, the form

$$\mathcal{F}_4(\sigma_1,\xi_1) = \dot{\xi}_1 \left(\dot{\sigma}_1 - \kappa^{-1}\dot{\xi}_1 \right).$$

Note that ξ_1 is now a vector $\begin{bmatrix} \xi & \dot{\xi} \end{bmatrix}$. Condition (3.2.16) implies that

$$\mathcal{F}_4\left(\sigma_1(t), \xi_1(t) \right) \ge 0.$$

Define $\tilde{\dot{\xi}} = i\omega\dot{\xi}$. Then the matrix $\Pi_4(\omega,\tau)$ is

$$\Pi_4(\omega,\tau) = -\omega^2 \operatorname{Re}\left[\kappa^{-1} + W(i\omega) \right].$$

Therefore, the frequency condition takes the following form. There exist nonnegative measures $\mu_j(\tau)$, $1 \le j \le 4$, such that for all real values of ω

$$\operatorname{Re}\left\{ Z(i\omega)\left[\kappa^{-1} + W(i\omega) \right] \right\} \ge \varepsilon > 0,$$

where

$$Z(i\omega) = \sum_{j=1}^{4} \mu_j(\tau) - i\omega \int_0^{+\infty} d\mu_2(\tau) + i\omega \int_0^{+\infty} d\mu_3(\tau) + \omega^2 \int_0^{+\infty} d\mu_4(\tau).$$

Set

$$\sum_{j=1}^{4} \mu_j(\tau) = 1;$$

$$\theta_1 = \int_0^{+\infty} d\mu_3(\tau) - \int_0^{+\infty} d\mu_2(\tau);$$

$$\theta_2 = \int_0^{+\infty} d\mu_4(\tau).$$

to obtain the frequency condition of the theorem.

Verification of minimal stability is given in the Subsect. 3.5.3.

By Lemma 2.6 the system is absolutely stable, which by definition means validity of the estimate (3.2.15).

By combining the arguments in the proof of this theorem with those from the previous two subsections we obtain two more results.

Theorem 3.14. *Assume the following*:

1) The linear block (3.1.1) *satisfies the regularity conditions;*
2) The functions $\alpha(t)$ *and* $K(t)$ *are absolutely continuous,* $K_0=0$*, and*

$$|\dot{\alpha}(\bullet)| \in L^2(0;\infty) ;$$

3) The function $\varphi(\sigma)$ *is differentiable and satisfies the slope restriction condition*

$$0 \leq \frac{\varphi(\sigma_1) - \varphi(\sigma_2)}{\sigma_1 - \sigma_2} \leq \kappa, \varphi(0) = 0 ;$$

4) There exist a constant θ_1*, a positive constant* θ_2*, and a nonnegative measure* $\mu(\tau)$*, such that* $\mu(\mathbb{R}) < 1$ *and for all real values of* ω

$$\mathrm{Re}\left\{\left[\kappa^{-1} + W(i\omega)\right]\left[1 + i\omega\theta_1 + \theta_2\omega^2 - \int_{-\infty}^{+\infty} e^{i\omega\tau} d\mu(\tau)\right]\right\} \geq \varepsilon > 0 . \quad (3.2.17)$$

Then for all functions $\sigma(\bullet)$ *and* $\xi(\bullet)$*, satisfying both* (3.1.1) *and* (3.1.2)*,* $\sigma(\bullet) \in L^2(0;+\infty)$ *and, furthermore, there exists a positive constant* λ*, independent of the function* $\alpha(\bullet)$*, such that*

$$\|\sigma(\bullet)\|^2 \leq \lambda\left[\|\alpha(\bullet)\|^2 + 2\int_0^{\sigma(0)} \varphi(\sigma)d\sigma + \zeta\frac{\sigma^2(0)}{2}\right]. \quad (3.2.18)$$

Theorem 3.15. *Assume that the hypotheses 1), 2), and 3) of the Theorem 3.14 are satisfied and the function* $\varphi(\sigma)$ *is odd. Suppose further that there exist a constant* θ_1*, a positive constant* θ_2*, and a signed measure* $\mu(\tau)$*, such that* $\mu(\mathbb{R}) < 1$ *and for all real values of* ω *the inequality* (3.2.17) *holds. Then the conclusion of Theorem 3.14 is true.*

3.3 Stability Multipliers for MIMO Systems

Consider now the system (3.1.1)-(3.1.2), but with $\sigma(t) \in \mathbb{R}^m$, $\alpha(t) \in \mathbb{R}^m$, $\xi(t) \in \mathbb{R}^m$ and K_0, $K(t)$ are $m \times m$ matrices.

The slope restriction condition will be replaced by an assumption that there exists a positive-definite (and hence nonsingular) matrix Λ such that for all $\sigma \in \mathbb{R}^m$

$$0 \le \frac{\partial \varphi(\sigma)}{\partial \sigma} < \Lambda \; ; \; \varphi(0) \equiv 0. \tag{3.3.1}$$

An example of a MIMO nonlinearity satisfying this condition is the case considered by Haddad and Kapila [57], in which each component of the function $\varphi(\sigma)$ satisfies the slope restriction condition with its own value of the constant κ. Then the Jacobian matrix is diagonal and finding the required matrix Λ is straightforward.

Here is an elementary example of the condition (3.3.1) for the function with a non diagonal Jacobian matrix. Let $\sigma = [x \quad y]^* \in \mathbb{R}^2$ and

$$\varphi(\sigma,t) = \begin{bmatrix} x + y \cos t \\ x \cos t + y \end{bmatrix}, \; \Lambda = \begin{bmatrix} \eta_1 & 0 \\ 0 & \eta_2 \end{bmatrix}.$$

Condition (3.3.1) is satisfied, provided that $2 < \eta_1 + \eta_2 < \eta_1 \eta_2$.

The stability criterion for systems with MIMO nonlinearities satisfying (3.3.1) is stated as follows:

Theorem 3.16. *Assume the following:*

1) The linear block (3.1.1) satisfies the regularity conditions;
2) The nonlinearity $\varphi(\sigma)$ is differentiable, satisfies condition (3.3.1), and

$$\left[\frac{\partial \varphi(\sigma)}{\partial \sigma} \right]^* - \frac{\partial \varphi(\sigma)}{\partial \sigma} \equiv 0 \; ; \tag{3.3.2}$$

3) There exists a nonnegative measure μ, such that $\mu(\mathbb{R}) < 1$ and for all real values of ω

$$\mathrm{Re}\left\{ \left[1 - \int_{-\infty}^{+\infty} e^{i\omega\tau} d\mu(\tau) \right] \left[\Lambda^{-1} + W(i\omega) \right] \right\} \ge \varepsilon I_m > 0. \tag{3.3.3}$$

Then for all functions $\sigma(\cdot)$ *and* $\xi(\cdot)$, *satisfying both* (3.1.1) *and* (3.1.2), $\sigma(\cdot) \in L^2(0;+\infty)$ *and, furthermore, there exists a positive constant* λ, *independent of the function* $\alpha(\cdot)$, *such that* $\|\sigma(\cdot)\| \le \lambda \|\alpha(\cdot)\|$.

Proof. First, similarly to the SISO case, define the quadratic form:

$$\mathcal{F}_1(\sigma_1, \xi_1) = \xi_1^*(\sigma_1 - \Lambda^{-1}\xi_1).$$

Condition (3.3.1) implies that

$$\mathcal{F}_1\big(\sigma(t), \xi(t)\big) \ge 0.$$

Indeed, by Mean Value Theorem, we have for some vector $\bar{\sigma}(t)$

$$\xi(t) = \frac{\partial \varphi(\sigma, t)}{\partial \sigma}\bigg|_{\sigma = \bar{\sigma}(t)} \sigma(t).$$

Because of (3.3.1), the matrix $\partial \varphi(\sigma, t) / \partial \sigma$ is positive-semidefinite and the matrix $I - \Lambda^{-1}[\partial \varphi(\sigma, t) / \partial \sigma]$ is positive-definite. Therefore, we have

$$\xi^*(t)[\sigma(t) - \Lambda^{-1}\xi(t)]$$

$$= \sigma^*(t)\left[\frac{\partial \varphi(\sigma, t)}{\partial \sigma}\bigg|_{\sigma = \bar{\sigma}(t)}\right]^*\left[I - \Lambda^{-1}\frac{\partial \varphi(\sigma, t)}{\partial \sigma}\bigg|_{\sigma = \bar{\sigma}(t)}\right]\sigma(t)$$

$$\ge 0.$$

Next, define two more quadratic forms:

$$\mathcal{F}_2(\sigma_1, \xi_1, \sigma_2, \xi_2) = (\sigma_1 - \Lambda^{-1}\xi_1)^*(\xi_1 - \xi_2);$$
$$\mathcal{F}_3(\sigma_1, \xi_1, \sigma_2, \xi_2) = \xi_1^*(\sigma_1 - \Lambda^{-1}\xi_1 - \sigma_2 + \Lambda^{-1}\xi_2).$$

Let us prove that for any $t_k > 0$

$$\int_0^{t_k} \mathcal{F}_j\big(\sigma(t), \xi(t), \sigma(t - \tau), \xi(t - \tau)\big)\, dt \ge 0, \quad j{=}2, 3. \tag{3.3.4}$$

Because of condition (3.3.2), we can define a function $\Phi(\sigma)$ as a path integral

$$\Phi(\sigma) = \int_0^{\sigma} \varphi(\sigma) d\sigma. \tag{3.3.5}$$

This function is convex since its Hessian is the Jacobian of the function $\varphi(\sigma)$, which is positive-semidefinite by (3.3.1). We can use this fact and apply Lemma 2.10 by setting $H(u,v) \equiv 0$; $x(t) = \varphi(\sigma_1(t))$, $g(x(t)) = \sigma(t) - \Lambda^{-1}\varphi(\sigma(t))$ for $j=2$ and $x(t) = \sigma(t) - \Lambda^{-1}\varphi(\sigma(t))$, $g(x(t)) = \varphi(\sigma_1(t))$ for $j=3$. It is easy to verify via Implicit Function Theorem that all the relevant functions are well defined and satisfy the conditions of Lemma 2.10. This proves (3.3.4).

The matrices $\Pi_j(\omega,\tau)$ are

$$\Pi_1(\omega,\tau) = -\text{Re}\left\{\Lambda^{-1} + W(i\omega)\right\};$$

$$\Pi_2(\omega,\tau) = -\text{Re}\left\{\left[\Lambda^{-1} + W(i\omega)\right]\left(1 - e^{i\omega\tau}\right)\right\};$$

$$\Pi_3(\omega,\tau) = -\text{Re}\left\{\left[\Lambda^{-1} + W(i\omega)\right]\left(1 - e^{-i\omega\tau}\right)\right\}.$$

Therefore, the FC takes the following form. There exist nonnegative measures μ_j, $j=1,2,3$, such that for all real values of ω

$$\text{Re}\left\{\left[\Lambda^{-1} + W(i\omega)\right]\left[\sum_{j=1}^{3}\mu_j(\mathbb{R}) - \int_0^{\infty}e^{i\omega\tau}d\mu_2(\tau) - \int_0^{\infty}e^{-i\omega\tau}d\mu_3(\tau)\right]\right\} \geq \varepsilon I_m > 0.$$

This is very similar to the FC for the SISO systems, except that instead of the scalar κ^{-1}, we now have a matrix Λ^{-1}. We now proceed in the same way by setting (without loss of generality)

$$\sum_{j=1}^{3}\mu_j(\mathbb{R}_+) = 1$$

and defining for any set $A \subset \mathbb{R}$ the measure $\mu(A) = \mu_2(A \cap \mathbb{R}_+) + \mu_3(-A \cap \mathbb{R}_+)$. Then, just as for the SISO case,

$$\int_0^{+\infty}e^{i\omega\tau}d\mu_2(\tau) + \int_0^{+\infty}e^{-i\omega\tau}d\mu_3(\tau)$$

$$= \int_0^{+\infty}e^{i\omega\tau}d\mu_2(\tau) - \int_{-\infty}^{0}e^{-i\omega\tau}d\mu_3(\tau)$$

$$= \int_{-\infty}^{+\infty}e^{i\omega\tau}d\mu(\tau).$$

Therefore, the frequency condition reduces to the one stated in hypothesis 3) of the theorem. The proof is completed by verification of minimal stability as described in Subsect. 3.5.3.

Just as for the SISO case, if the nonlinearity is odd, we can weaken the requirement that the measure μ must be nonnegative.

Theorem 3.17. *Assume that the conditions 1) and 2) of Theorem 3.16 hold and, in addition, $\varphi(-\sigma) \equiv \varphi(\sigma)$. Assume further that there exists a signed finite measure μ, such that $\mu(\mathbb{R}) < 1$ and for all real values of ω*

$$\mathrm{Re}\left\{\left[1 - \int_{-\infty}^{+\infty} e^{i\omega\tau} d\mu(\tau)\right]\left[\Lambda^{-1} + W(i\omega)\right]\right\} \geq \varepsilon I_m > 0. \qquad (3.3.6)$$

Then the conclusion of Theorem 3.16 holds.

Proof. Just as in the SISO case, we can define, in addition to the quadratic forms $\mathcal{F}_j(\sigma_1, \xi_1, \sigma_2, \xi_2)$, $j=1,2,3$, two more forms:

$$\mathcal{F}_4(\sigma_1, \xi_1, \sigma_2, \xi_2) = (\sigma_1 - \Lambda^{-1}\xi_1)^*(\xi_1 + \xi_2);$$
$$\mathcal{F}_5(\sigma_1, \xi_1, \sigma_2, \xi_2) = \xi_1^*(\sigma_1 - \Lambda^{-1}\xi_1 + \sigma_2 - \Lambda^{-1}\xi_2).$$

Let us prove that for any $t_k > 0$

$$\int_0^{t_k} \mathcal{F}_j\left(\sigma(t), \xi(t), \sigma(t-\tau), \xi(t-\tau)\right) dt \geq 0, \quad j=4, 5. \qquad (3.3.7)$$

Similarly to the proof of Theorem 3.16, define the function $\Phi(\sigma)$ by (3.3.5). By the same argument as in the proof of Theorem 3.16, this function is convex. We can now apply the second part of Lemma 2.10 by setting $H(u,v) \equiv 0$; $x(t) = \varphi(\sigma_1(t))$, $g(x(t)) = \sigma(t) - \Lambda^{-1}\varphi(\sigma(t))$ for $j=4$ and $x(t) = \sigma(t) - \Lambda^{-1}\varphi(\sigma(t))$, $g(x(t)) = \varphi(\sigma_1(t))$ for $j=5$.

The matrices $\Pi_j(\omega, \tau)$ are:

$$\Pi_4(\omega, \tau) = -\mathrm{Re}\left\{\left[\Lambda^{-1} + W(i\omega)\right]\left(1 + e^{-i\omega\tau}\right)\right\};$$
$$\Pi_5(\omega, \tau) = -\mathrm{Re}\left\{\left[\Lambda^{-1} + W(i\omega)\right]\left(1 + e^{i\omega\tau}\right)\right\}.$$

Similarly to the SISO case, the frequency condition now takes the following form. There exist nonnegative measures $\mu_j(\tau), 1 \le j \le 5$, such that for all real values of ω

$$\mathrm{Re}\left\{\left[\Lambda^{-1} + W(i\omega)\right]Z(i\omega)\right\} \ge \varepsilon I_m > 0.$$

The multiplier $Z(i\omega)$ is given by

$$Z(i\omega) \triangleq \sum_{j=1}^{5} \mu_j\left(\mathbb{R}_+\right) - \int_0^{+\infty} e^{i\omega\tau} d\mu_2(\tau) - \int_0^{+\infty} e^{-i\omega\tau} d\mu_3(\tau)$$
$$+ \int_0^{+\infty} e^{-i\omega\tau} d\mu_4(\tau) + \int_0^{+\infty} e^{i\omega\tau} d\mu_5(\tau).$$

Again, without loss of generality we can set

$$\sum_{j=1}^{5} \mu_j\left(\mathbb{R}_+\right) = 1.$$

Now, just as in the SISO case, define for any set $A \subset \mathbb{R}$ two signed measures $\overline{\mu}_2(A) = \mu_2(A) - \mu_4(A)$ and $\overline{\mu}_3(A) = \mu_3(A) - \mu_5(A)$ and then the measure

$$\mu(A) = \overline{\mu}_2(A \cap \mathbb{R}_+) + \overline{\mu}_3(-A \cap \mathbb{R}_+).$$

The proof is completed by verifying minimal stability as described in Subsect. 3.5.3.

We have thus proved the MIMO analogues of Theorems 3.8 and 3.9. Interestingly, Zames and Falb in their original paper stated that their result, proved for SISO nonlinearities, applies directly to MIMO systems. This turned out to be one of those "easy to see" statements that ultimately proved to be incorrect as was shown by Safonov and Kulkarni [128]. As we have seen, the method of delay-integral-quadratic constraints applies to both SISO and MIMO systems with relatively little difference.

Again, similarly to the SISO case, we combine the arguments in the proofs of Theorems 3.16 and 3.17 with those used in the proof of Theorem 3.13. We obtain the following two results.

Theorem 3.18. *Assume that the conditions 1) and 2) of Theorem 3.16 hold and, in addition: functions $\alpha(t)$ and $K(t)$ are absolutely continuous, $K_0 = 0$, and $|\dot{\alpha}(\bullet)| \in L^2(0;\infty)$. Assume further that there exists a nonnegative finite measure μ, such that $\mu(\mathbb{R}) < 1$ and for all real values of ω*

$$\mathrm{Re}\left\{\left[1+\vartheta\omega^2 - \int_{-\infty}^{+\infty} e^{i\omega\tau} d\mu(\tau)\right]\left[\Lambda^{-1} + W(i\omega)\right]\right\} \geq \varepsilon I_m > 0. \qquad (3.3.8)$$

Then the conclusion of Theorem 3.16 holds.

Theorem 3.19. *Assume that the conditions 1) and 2) of the Theorem 3.16 hold and, in addition: $\varphi(-\sigma) \equiv \varphi(\sigma)$, functions $\alpha(t)$ and $K(t)$ are absolutely continuous, $K_0 = 0$, and $|\dot\alpha(\cdot)| \in L^2(0; \infty)$. Assume further that there exists a signed finite measure μ, such that $\mu(\mathbb{R}) < 1$ and for all real values of ω*

$$\mathrm{Re}\left\{\left[1+\vartheta\omega^2 - \int_{-\infty}^{+\infty} e^{i\omega\tau} d\mu(\tau)\right]\left[\Lambda^{-1} + W(i\omega)\right]\right\} \geq \varepsilon I_m > 0. \qquad (3.3.9)$$

Then the conclusion of the Theorem 3.16 holds.

The proof of both of these theorems proceeds in a manner similar to Theorems 3.14 and 3.15 except that we first, using the fact that the matrix $K(t)$ is absolutely continuous, augment the linear block by adding the equation:

$$\dot\sigma(t) = \dot\alpha(t) + \int_0^t \frac{\partial K(t-s)}{\partial t}\xi(t)ds + K(0)\xi(t).$$

The nonlinear block is augmented by adding the equation

$$\dot\xi(t) = \frac{\partial\varphi(\sigma)}{\partial\sigma}\dot\sigma(t).$$

Then we define the following additional quadratic form:

$$\mathcal{F}_6\left(\sigma_1, \xi_1, \dot\sigma_1, \dot\xi_1\right) = \dot\xi_1^*\left(\dot\sigma_1 - \Lambda^{-1}\dot\xi_1\right).$$

Condition (3.3.1) implies that $\mathcal{F}_6\left(\sigma_1(t), \xi_1(t), \dot\sigma_1(t), \dot\xi_1(t)\right) \geq 0$. Indeed, recall that the matrix $\partial\varphi(\sigma, t)/\partial\sigma$ is positive-semidefinite and the matrix $I - \Lambda^{-1}[\partial\varphi(\sigma, t)/\partial\sigma]$ is positive-definite. Therefore, we have

$$\dot\xi^*(t)[\dot\sigma(t) - \Lambda^{-1}\dot\xi(t)]$$
$$= \dot\sigma^*(t)\left[\frac{\partial\varphi(\sigma, t)}{\partial\sigma}\right]^*\left[I - \Lambda^{-1}\frac{\partial\varphi(\sigma, t)}{\partial\sigma}\right]\dot\sigma(t)$$
$$\geq 0.$$

We can now apply similar arguments as in the proofs of Theorems 3.16 and 3.17.

These results can be used to develop an approach to solving an output regulation problem. Consider the system:

$$\dot{x} = Ax + u(x) + \phi(x). \tag{3.3.10}$$

Here $x(t) \in \mathbb{R}^m$. We assume that the matrix A is Hurwitz stable and that the function $\phi(x)$ is differentiable and satisfies the condition (3.3.1). We can rewrite this system in the form (3.1.1)-(3.1.2) by setting $\sigma(t) = x(t)$, $K(t) = e^{At}$, $\alpha(t) = e^{At} x(0)$, and $K_0 = 0$. Our objective is to find a control law $u(x)$, which guarantees absolute stability of (3.3.10). This is a variation of the output regulation problem described in the book [114]. Theorem 3.18 suggests an approach to solving this problem.

Define the vector function $\varphi(x) = u(x) + \phi(x)$ and the transfer matrix

$$W(s) = [sI - A]^{-1}.$$

It follows from Theorem 3.18 that the answer to this problem is given by the control law $u(x)$ satisfying the following conditions:

1.   The vector function $u(x)$ is differentiable and

$$\left[\frac{\partial\left(u(x) + \phi(x)\right)}{\partial x}\right]^* = \frac{\partial\left(u(x) + \phi(x)\right)}{\partial x}; \tag{3.3.11}$$

2.   There exists a constant matrix U such that

$$0 \le \frac{\partial u}{\partial x} \le U; \tag{3.3.12}$$

3.   There exists a nonnegative finite measure μ, such that $\mu(\mathbb{R}) < 1$ and for all real values of ω

$$\mathrm{Re}\left\{Z(i\omega)\left[(\Lambda + U)^{-1} + W(i\omega)\right]\right\} \ge \varepsilon I_m > 0, \tag{3.3.13}$$

where

$$Z(i\omega) = 1 + \vartheta\omega^2 - \int_{\mathbb{R}} e^{i\omega\tau} d\mu(\tau).$$

Note that we did not require the Jacobian of the function $\phi(x)$ to be symmetric. Instead, this requirement is imposed on the function $\varphi(x) = u(x) + \phi(x)$ in order to apply Theorem 3.18.

Finding the desired controller may be accomplished in two steps. First, the FC (3.3.13) can be used (with a suitably chosen multiplier) to find the matrix U. Then, the conditions (3.3.11) and (3.3.12) can be used to choose a function $u(x)$, which is obviously not unique.

3.4   Constructive Stability Criteria Based on Multipliers

Results presented thus far in this chapter are very general. However, they also are difficult to verify, because they involve selecting some arbitrary functions. For this reason, many researchers gave a considerable amount of attention to the problem of constructing the multipliers. Methods, generally, fall into three large groups: Algebraic, geometric, and numerical.

Algebraic criteria focus on finding explicit closed-form expressions for stability multipliers, which replaces finding of an arbitrary function with selection of suitable parameters. Most of the algebraic criteria have already been discussed as part of the historical survey in Chapter 1. In this section we focus on geometric interpretation and numerical implementation of stability multipliers.

Each of these two approaches has some advantages and drawbacks. Geometric approach is applicable only to SISO nonlinearities but does not impose any restrictions on the transfer function of the linear block. To contrast, numerical methods can be used for MIMO nonlinearities, but have a built-in assumption that the linear block has a matrix realization.

3.4.1   Geometric Interpretation of Stability Multipliers

All frequency conditions in this chapter have the general form:

$$\mathrm{Re}\left\{Z(i\omega)\left[\kappa^{-1} + W(i\omega)\right]\right\} \geq \varepsilon > 0. \tag{3.4.1}$$

This can be rewritten (with a slight abuse of notation) as

$$\left[\kappa^{-1} + \mathrm{Re}\,W(i\omega)\right]\mathrm{Re}\,Z(i\omega) - \mathrm{Im}\,W(i\omega)\,\mathrm{Im}\,Z(i\omega) > 0. \tag{3.4.2}$$

In order for (3.4.1) and (3.4.2) to be equivalent, we must require that (3.4.2) holds at infinity.

Define two functions:

$$\Phi(\omega) = \frac{\kappa^{-1} + \mathrm{Re}\,W(i\omega)}{\mathrm{Im}\,W(i\omega)}, \ \Psi(\omega) = \frac{\mathrm{Im}\,Z(i\omega)}{\mathrm{Re}\,Z(i\omega)}. \tag{3.4.3}$$

Note that the function $\Phi(\omega)$ involves only the frequency response of the linear block, while the function $\Psi(\omega)$ involves only the multiplier. Inequality (3.4.2) holds if and only if the graph of the function $\Psi(\omega)$ lies above the graph of the function $\Phi(\omega)$ on the intervals where $\operatorname{Im} W(i\omega) < 0$ and below the graph of the function $\Phi(\omega)$ on the intervals where $\operatorname{Im} W(i\omega) > 0$.

Note that, since the feedback system is stable for $\varphi(\sigma,t) = k\sigma$ with $0 \leq k \leq \kappa$, the Nyquist plot of the function $W(i\omega)$ does not encircle the point $(-\kappa^{-1}, 0)$. Therefore, the numerator of the function $\Phi(\omega)$ is positive whenever $\operatorname{Im} W(i\omega) = 0$. This implies that the graph of the function $\Phi(\omega)$ consists of branches with asymptotes defined by $\operatorname{Im} W(i\omega) = 0$. Both ends of each branch tend to either $+\infty$ (they will be called "stalactites") or to $-\infty$ (they will be called "stalagmites"). Therefore, we can now state that (3.4.2) holds if and only if the graph of the function $\Psi(\omega)$ separates stalactites from stalagmites. These concepts were first introduced by Lipatov [88, 90, 91]. For these reasons we call the plots of the functions $\Phi(\omega)$ and $\Psi(\omega)$ "the Lipatov plots." Clearly, each multiplier will have its own Lipatov plot.

The shapes of the curves $\Psi(\omega)$ for some specific stability criteria are described in Table 3.1.

Table 3.1 Lipatov plots for some stability criteria

Criterion	$\Psi(\omega)$	*Shape of* $\Psi(\omega)$
Circle	$\Psi(\omega) \equiv 0$	Abscissa axis
Popov	$\Psi(\omega) = \theta\omega$	Oblique straight line passing through the origin
Off-axis circle/Voronov	$\Psi(\omega) = \theta$	Horizontal straight line
Yakubovich	$\Psi(\omega) = \dfrac{\theta_1\omega}{1 + \theta_2\omega^2}$	Parabola passing through the origin

If none of the simple stability criteria from Table 3.1 gives a satisfactory result, then in order to use these geometric concepts it is necessary to draw a curve $\Psi(\omega)$ in such a way that stalactites are separated from the stalagmites and then to verify the condition

$$\int_0^\infty |z(t)| \, dt < 1 \tag{3.4.4}$$

for the inverse Fourier transform $z(t)$ of the resulting multiplier. For the sake of simplicity we limit ourselves to cases when $z(t)$ is absolutely continuous. The most common approach is to use one of the simple multipliers given in Table 3.1 and then modify the resulting curve to make sure that the required separation is achieved.

Verification of (3.4.4) can be difficult. To simplify this task, Lipatov [90, 91], Freedman [49], and Barabanov [19] considered only the multipliers with $\operatorname{Re} Z(i\omega) \equiv 1$ and set $\operatorname{Im} Z(i\omega) = \theta\omega + V(\omega)$. They imposed a further simplification by setting $\theta = 0$, i.e., drop the "Popov term" from the multiplier.

Let $v(t)$ be the inverse Fourier transform of $V(\omega)$. Barabanov [19] proved the estimate:

$$\int_0^\infty |v(t)| \, dt \le \sqrt{2} \left[\int_{-\infty}^{+\infty} [V(\omega)]^2 d\omega \right]^{1/4} \left[\int_{-\infty}^{+\infty} [V'(\omega)]^2 d\omega \right]^{1/4}.$$

Hence, verification of (3.4.4) can be replaced with verification of

$$\int_{-\infty}^{+\infty} [V(\omega)]^2 d\omega \int_{-\infty}^{+\infty} [V'(\omega)]^2 d\omega < \frac{1}{4}. \tag{3.4.5}$$

The latter is generally an easier task than verification of (3.4.4).

Freedman [49] used a somewhat different approach. Instead of (3.4.3), he defined the function $\Phi(\omega)$:

$$\Phi(\omega) = \begin{bmatrix} \pi/2 - \arg\left[\kappa^{-1} + W(i\omega) \right] & \text{if } \arg\left[\kappa^{-1} + W(i\omega) \right] \ge 0 \\ -\pi/2 - \arg\left[\kappa^{-1} + W(i\omega) \right] & \text{if } \arg\left[\kappa^{-1} + W(i\omega) \right] \le 0 \end{bmatrix}.$$

It can be proved that graph of this function has the same geometric properties as the graph of the function $\Phi(\omega)$ defined by (3.4.3). Freedman then defined the function $s(\omega)$ as follows:

$$s(\omega) = \begin{bmatrix} \Phi(\omega) & \text{if } \arg\left[\kappa^{-1} + W(i\omega) \right] \ge \pi/2 \\ 0 & \text{if } \arg\left[\kappa^{-1} + W(i\omega) \right] \le \pi/2 \end{bmatrix}.$$

The frequency condition is then proved to take the form:

$$\int_{-\infty}^{+\infty}[s(\omega)]^2 d\omega \int_{-\infty}^{+\infty}[s'(\omega)]^2 d\omega < \frac{\ln^4 2}{4}. \tag{3.4.6}$$

Barabanov [19] showed that the estimate (3.4.5) gives a better stability result than (3.4.6).

In papers [89, 92, 93], Lipatov and his coworkers also developed a geometric procedure using the Nyquist plot directly. However, these methods require measuring some angles on the plot, which might explain why they have not gained much acceptance in the control engineering community.

In [20], Barabanov used the above-described geometric concepts to derive the following result. Let $f(z)$ be a function of a complex variable, analytic in the right half-plane and also in the right half-plane $f(z) \neq ia$ for any real a with absolute value greater than 1. Then the system is absolutely stable if there exists a function $f(z)$ satisfying these two properties, such that for all real values of ω

$$\kappa^{-1} + \operatorname{Re} W(i\omega) > f(\omega)\operatorname{Im} W(i\omega).$$

It may be worth noting that for stationary nonlinearities, this geometric interpretation involves some "coarsening" of the frequency conditions by dropping the "Popov term." This term is not present in the case of semimonotone (as well as other nonstationary) nonlinearities, which reduces somewhat the coarsening.

3.4.2 Numerical Implementation of Stability Multipliers

As mentioned in Sect. 1.3, frequency-domain criteria were developed partly because there were no effective numerical methods for solving linear matrix inequalities (LMIs). At the present time such methods have been developed (see, for example, the monograph [33]). For this reason, many researchers now prefer to express stability criteria in the form of the LMIs.

The relationship between stability multipliers and the LMIs is mentioned in passing on page 125 of [33]. Let us discuss it in some more detail for the case of stationary MIMO nonlinearities, i.e., when the system (3.1.1)-(3.1.2) with the vector function $\varphi(\sigma)$ satisfies (3.3.1).

Let $\{A, B, C, D\}$ be a matrix realization of $\Lambda^{-1} + W(s)$ or $\kappa^{-1} + W(s)$, depending on whether a SISO or a MIMO nonlinearity is considered, and let $\{A_m, B_m, C_m, D_m\}$ be a matrix realization of the multiplier $Z(s)$. Define matrices:

$$\bar{A} = \begin{bmatrix} A & 0 \\ B_m C & A_m \end{bmatrix},\ \bar{B} = \begin{bmatrix} B \\ B_m D \end{bmatrix},\ \bar{C} = \begin{bmatrix} D_m C & C_m \end{bmatrix},\ \bar{D} = D_m D.$$

Then the frequency condition for stability is equivalent to the following LMI in the matrices $P>0$ and the ones from the realization of the multiplier:

$$\begin{bmatrix} \overline{A}*P+P\overline{A} & P\overline{B}+\overline{C}* \\ \overline{B}*P+\overline{C} & \overline{D}+\overline{D}* \end{bmatrix} < 0.$$

Another approach to numerical implementation of stability multipliers was proposed by Safonov and Wyetzner [127]. They wrote the stability multiplier in the form:

$$M_n(i\omega) = \frac{1+i\omega\theta}{1+i\omega/n}.$$

Note that the Popov criterion is obtained in the limit as $n \to \infty$.

The problem of finding the multiplier is then reduced to finding the maximum value of the constant ρ, subject to inequality constraints:

$$\forall \omega: \rho + \mathrm{Re}\{W(i\omega)[1-M_1(i\omega)]\} \le 0;$$
$$\forall t: m_1(t) \ge 0,$$

where $m_1(t)$ is the inverse Fourier transform of $M_1(\omega)$.

This is an infinite-dimensional linear programming problem. Safonov and Wyetzner then proposed a method to discretize it and solve it numerically. They proved that the resulting stability criterion is always at least as strong as that of Popov. Subsequently, Safonov with two other coworkers Mancera and Chang [37, 38] improved this computational algorithm.

There are some other approaches to numerical implementation of stability multipliers that also use some optimization methods. One of them is described in two papers by Skorodinskii [132, 133]. Another can be found in the work of Gapski and Geromel [51]. One of the more recent papers that consider this subject is the presentation by Turner et al [136].

3.5 Proofs of Lemmas and Verification of Minimal Stability

3.5.1 Proof of Lemma 3.1

An Additional Lemma

Proof of Lemma 3.1 relies on the following additional lemma.

Lemma 3.20. *Suppose that the function* $\varphi(\sigma,t)$ *satisfies the conditions*

$$0 \le \frac{\varphi(\sigma_1,t)-\varphi(\sigma_2,t)}{\sigma_1-\sigma_2} \le \kappa, \varphi(0,t) \equiv 0 \tag{3.5.1}$$

and

$$\sigma[\varphi(\sigma,t-\tau)-\varphi(\sigma,t)]\geq 0. \qquad (3.5.2)$$

Let $\chi(x,t)$ be the solution of the equation $x=\kappa\chi-\varphi(\chi,t)$. Then the func-
tion $g(x,t)\triangleq\varphi(\chi(x,t),t)$ satisfies the condition (3.5.2) for the same values of τ as
the function $\varphi(\sigma,t)$.

Proof. From the defining equation for the function $g(x,t)$,

$$x[g(x,t-\tau)-g(x,t)]=x[\varphi(\chi(x,t-\tau),t-\tau)-\varphi(\chi(x,t),t)].$$

Because of (3.5.1), x and $\chi(x,t)$ have the same sign. Therefore, the inequality
(3.5.2) implies

$$x[\varphi(\chi(x,t-\tau),t-\tau)-\varphi(\chi(x,t-\tau),t)]\geq 0.$$

This implies that the conclusion of the lemma is true if the following inequality
holds:

$$x[\varphi(\chi(x,t-\tau),t)-\varphi(\chi(x,t),t)]\geq 0. \qquad (3.5.3)$$

Let us now prove (3.5.3). If $\chi(x,t-\tau)=\chi(x,t)$, then it is certainly true. Therefore, assume that $\chi(x,t-\tau)\neq\chi(x,t)$. In this case, (3.5.3) will follow from the fact
that the function $\varphi(\sigma,t)$ is nondecreasing in σ if we prove the inequality:

$$x[\chi(x,t-\tau)-\chi(x,t)]>0. \qquad (3.5.4)$$

We have the following two equations:

$$x=\kappa\chi(x,t)-\varphi(\chi(x,t),t);$$
$$x=\kappa\chi(x,t-\tau)-\varphi(\chi(x,t-\tau),t-\tau).$$

Subtracting these two equations, we obtain

$$\kappa[\chi(x,t-\tau)-\chi(x,t)]-[\varphi(\chi(x,t-\tau),t-\tau)-\varphi(\chi(x,t),t)]=0.$$

By adding and subtracting $\varphi(\chi(x,t-\tau),t)$, we can rewrite this equation as follows:

$$\kappa[\chi(x,t-\tau)-\chi(x,t)]-[\varphi(\chi(x,t-\tau),t)-\varphi(\chi(x,t),t)]$$
$$-[\varphi(\chi(x,t-\tau),t-\tau)-\varphi(\chi(x,t-\tau),t)]=0.$$

Multiply both sides by x. Since $\chi(x,t-\tau)\neq\chi(x,t)$, we have

$$\left[\kappa-\frac{\varphi(\chi(x,t-\tau),t)-\varphi(\chi(x,t),t)}{\chi(x,t-\tau)-\chi(x,t)}\right][\chi(x,t-\tau)-\chi(x,t)]x$$
$$-[\varphi(\chi(x,t-\tau),t-\tau)-\varphi(\chi(x,t-\tau),t)]x=0.$$

From (3.5.2) we have

$$[\varphi(\chi(x,t-\tau),t-\tau)-\varphi(\chi(x,t-\tau),t)]x\geq0.$$

From (3.5.1) we have

$$\kappa-\frac{\varphi(\chi(x,t-\tau),t)-\varphi(\chi(x,t),t)}{\chi(x,t-\tau)-\chi(x,t)}>0.$$

Therefore, $x[\chi(x,t-\tau)-\chi(x,t)]>0$ as required.

Let us illustrate this lemma for a simple case when the function $\varphi(\sigma,t)$ is differentiable in each variable, nonincreasing in t for positive σ and nondecreasing in t for negative σ. Then differentiation of the equation $x=\kappa\chi-\varphi(\chi,t)$ gives

$$\left(\kappa-\frac{\partial\varphi}{\partial\chi}\right)\frac{d\chi}{dt}-\frac{\partial\varphi}{\partial t}=0.$$

We can use this equation to compute the derivative of the function $g(x,t)$:

$$\frac{\partial g}{\partial t}=\frac{\partial\varphi}{\partial\chi}\frac{d\chi}{dt}+\frac{\partial\varphi}{\partial t}=\left(1+\frac{\dfrac{\partial\varphi}{\partial\chi}}{\kappa-\dfrac{\partial\varphi}{\partial\chi}}\right)\frac{\partial\varphi}{\partial t}.$$

The condition (3.5.1) implies that the expression in parentheses is positive. There-
fore, the derivative has the same sign as $\partial\varphi/\partial t$. The signs of x and χ are the same.
Therefore, the function $g(x,t)$ satisfies the conclusion of the lemma.

Proof of Lemma 3.1

Let us now proceed with proof of Lemma 3.1. First, let us consider the case when
the following inequality is true:

$$\frac{\Delta\varphi(\sigma,t)}{\Delta t} \leq \kappa + \varepsilon, \tag{3.5.5}$$

where ε is a small number.

Define the function $g(x,t)$ in the same way as in Lemma 3.20. This function satisfies the hypotheses 1)-3) and, by Lemma 3.20, the assumption 4) of Lemma
2.10. Let us verify that it also satisfies the assumption 5) of Lemma 2.10
with $H(u,v) \equiv 0$, which is equivalent to stating that it is nondecreasing in the
variable x.

Consider the following two equations:

$$x = \kappa\chi(x,t) - \varphi(\chi(x,t),t);$$
$$y = \kappa\chi(y,t) - \varphi(\chi(y,t),t).$$

Subtracting, we obtain

$$x - y = \kappa[\chi(x,t) - \chi(y,t)] - [\varphi(\chi(x,t),t) - \varphi(\chi(y,t),t)]. \tag{3.5.6}$$

If $\chi(x,t) - \chi(y,t) = 0$, then $x=y$. Otherwise, (3.5.6) can be rewritten in the form

$$x - y = \left[\kappa - \frac{\varphi(\chi(x,t),t) - \varphi(\chi(y,t),t)}{\chi(x,t) - \chi(y,t)}\right][\chi(x,t) - \chi(y,t)].$$

The condition (3.5.1) implies that the first factor in the right-hand side is positive.
Hence, $\chi(x,t) - \chi(y,t)$ has the same sign as x-y. Since the function $\varphi(\sigma,t)$ is
nondecreasing in σ, the difference $g(x,t) - g(y,t)$ also has the same sign as x-y.
Therefore, the function $g(x,t)$ is nondecreasing in x.

It is thus proved that the function $g(x,t)$ satisfies all the conditions of
Lemma 2.10. Hence, denoting $x(t) = \eta(t) \triangleq \kappa\sigma(t) - \varphi(\sigma(t),t)$, we have the
inequalities in the conclusion of Lemma 3.1.

Now let us consider the more general case

$$\frac{\Delta\varphi(\sigma,t)}{\Delta t} \leq \kappa.$$

Define the new function $\varphi_\varepsilon(\sigma,t) = \varphi(\sigma,t) - \varepsilon\sigma$. The conclusion of Lemma 3.1 holds for this function if $\eta(t)$ is replaced with $\eta_\varepsilon(t) \triangleq \kappa\sigma - \varphi_\varepsilon(\sigma,t)$. In the limit as $\varepsilon \to 0$ we find that it holds for the functions $\varphi(\sigma,t)$ and $\eta(t)$ as well. The proof is complete.

3.5.2 Proof of Lemma 3.2

An Additional Lemma

Proof of Lemma 3.2 relies on the following lemma.

Lemma 3.21. *Assume that the function $\varphi(\sigma,t)$ satisfies the conditions:*

$$0 \leq \frac{\varphi(\sigma_1,t) - \varphi(\sigma_2,t)}{\sigma_1 - \sigma_2}, \varphi(0,t) \equiv 0. \qquad (3.5.7)$$

Suppose that there exists a number τ such that for all x and t

$$\sigma\big[\varphi(\sigma,t-\tau) - \varphi(\sigma,t)\big] \leq 0. \qquad (3.5.8)$$

Define $a(t) = \inf_\sigma \varphi(\sigma,t)$, $b(t) = \sup_\sigma \varphi(\sigma,t)$. Let $\chi(x,t)$ be the root of the equation on the interval $\big(a(t); b(t)\big)$. Then for all x and t

$$x\big[\chi(x,t-\tau) - \chi(x,t)\big] \geq 0.$$

Proof. If $\chi(x,t-\tau) = \chi(x,t)$, then the conclusion of the lemma is obviously true. Therefore, assume that $\chi(x,t-\tau) \neq \chi(x,t)$. We have two equations:

$$x = \varphi\big(\chi(x,t),t\big);$$
$$x = \varphi\big(\chi(x,t-\tau),t-\tau\big).$$

Subtracting, we obtain

$$\varphi(\chi(x,t-\tau),t-\tau)-\varphi(\chi(x,t),t)=0.$$

Now add to and subtract from the left-hand side the expression $\varphi(\chi(x,t-\tau),t-\tau)$. The equation takes the form

$$\left[\varphi(\chi(x,t-\tau),t)-\varphi(\chi(x,t),t)\right]+\left[\varphi(\chi(x,t-\tau),t-\tau)-\varphi(\chi(x,t-\tau),t)\right]=0.$$

Since we have assumed that $\chi(x,t-\tau)\neq\chi(x,t)$, we can rewrite this equation as follows, after multiplying both sides by x:

$$\frac{\varphi(\chi(x,t-\tau),t)-\varphi(\chi(x,t),t)}{\chi(x,t-\tau)-\chi(x,t)}\left[\chi(x,t-\tau)-\chi(x,t)\right]x$$
$$+\left[\varphi(\chi(x,t-\tau),t-\tau)-\varphi(\chi(x,t-\tau),t)\right]x=0.$$

Condition (3.5.7) implies

$$\frac{\varphi(\chi(x,t-\tau),t)-\varphi(\chi(x,t),t)}{\chi(x,t-\tau)-\chi(x,t)}>0.$$

On the other hand, condition (3.5.8) implies

$$\chi(x,t-\tau)\left[\varphi(\chi(x,t-\tau),t-\tau)-\varphi(\chi(x,t-\tau),t)\right]<0.$$

Therefore, $x\left[\chi(x,t-\tau)-\chi(x,t)\right]\geq0$, which proves the lemma.

Let us illustrate this lemma with a simple example. Condition (3.5.8) is clearly true if the function $\varphi(\sigma,t)$ is nondecreasing in t for positive σ and nonincreasing in t for negative σ. Assume further that this function is differentiable with respect to each variable. Differentiating the equation $x=\varphi(\chi(t),t)$ yields

$$\frac{\partial\varphi}{\partial\chi}\frac{\partial\chi}{\partial t}+\frac{\partial\varphi}{\partial t}=0.$$

This implies that the signs of the derivatives $\partial\chi/\partial t$ and $\partial\varphi/\partial t$ are opposite. Therefore, the function $\chi(x,t)$ is nonincreasing in t for positive x and nondecreasing in t for negative x, i.e., it satisfies the conclusion of the lemma.

Proof of Lemma 3.2

Let us now proceed with the proof of Lemma 3.2.

First, let us consider a simpler case with the inequality

$$0 < \varepsilon \le \frac{\varphi(\sigma_1,t)-\varphi(\sigma_2,t)}{\sigma_1-\sigma_2} \le \kappa, \varphi(0,t) \equiv 0. \tag{3.5.9}$$

Define the function $g(x,t) = \chi(x,t) - \kappa^{-1}x$, where $\chi(x,t)$ is the function defined in the statement of lemma 3.18. It follows from Lemma 3.21 and the inequality (3.5.9) that this function satisfies condition 4) of lemma 2.10. Furthermore, it is nondecreasing in the variable σ and hence satisfies condition 5) of Lemma 2.10. Conclusion of Lemma 3.2 now follows from the application of Lemma 2.10 to the function $g\big(x(t),t\big) = \sigma(t) - \kappa^{-1}\varphi\big(\sigma(t),t\big)$ with $x(t) = \xi(t)$.

For the more general case of the inequality

$$0 \le \frac{\varphi(\sigma_1,t)-\varphi(\sigma_2,t)}{\sigma_1-\sigma_2}.$$

we define the function $\varphi_\varepsilon(\sigma,t) + \varepsilon\sigma$. This function satisfies the condition

$$0 < \varepsilon \le \frac{\varphi(\sigma_1,t)-\varphi(\sigma_2,t)}{\sigma_1-\sigma_2} \le \kappa + \varepsilon, \varphi(0,t) \equiv 0.$$

Now apply Lemma 2.10 to the function $g(x(t),t) = \sigma(t) - (\kappa+\varepsilon)^{-1}\varphi_\varepsilon(\sigma(t),t)$ with $x(t) = \xi(t)$. In the limit as $\varepsilon \to 0$, we obtain the conclusion of Lemma 3.2. The proof is complete.

3.5.3 Verification of Minimal Stability

Verification of minimal stability is the same for all the theorems in this Chapter. For any process $z(\bullet)$ we construct its bounded continuation as follows. Let $t_k \to +\infty$ be an arbitrary sequence. Define $m_k = \max_{t\in[0;t_k]} |\sigma(t)|$ and

$$\varphi_k(\sigma,t) = \begin{cases} \varphi(\sigma,t) \text{ if } |\sigma| \le m_k \\ \varphi(m_k,t) \text{ if } |\sigma| > m_k \\ \varphi(-m_k,t) \text{ if } |\sigma| < -m_k \end{cases}.$$

If the function $\varphi(\sigma,t)$ depends only on the variable σ, then the functions $\varphi_k(\sigma,t)$ are defined in the same way. If the conditions of the appropriate theorem require the function $\varphi(\sigma,t)$ to be differentiable in σ, we replace the functions $\varphi_k(\sigma,t)$ with their differentiable approximations.

All the constraints used in the proofs of the theorems of this chapter hold for the process $z_k(\bullet)$ defined as a solution of the system consisting of (3.1.1) and

$$\xi(t) = \varphi_k(\sigma(t), t) .$$

Therefore, the process $z_k(\bullet)$ is a bounded continuation of the process $z(\bullet)$. By Theorem 2.9 it is a stable continuation and since this construction is applicable to any process, the system is minimally stable.

Chapter 4
Time-Periodic Systems

4.1 Frequency Condition for Time Periodic Systems

In this chapter we consider a system with the linear block given by the Volterra integral equation:

$$\sigma(t) = \alpha(t) + K_0\xi(t) + \int_0^t K(t-s)\xi(s)ds .\tag{4.1.1}$$

The nonlinear block will be assumed to be represented by the equation

$$\xi = \varphi(\sigma,t) .\tag{4.1.2}$$

As before, $\sigma(t) \in \mathbb{R}^m$; $\xi(t) \in \mathbb{R}^p$; and K_0, $K(t)$ are matrices of appropriate dimensions. Throughout the chapter it will be assumed that $\varphi(\sigma,t+T) = \varphi(\sigma,t)$. Additional assumptions concerning this function will be stated in hypotheses of the theorems.

The results of this chapter, similarly to Chap. 3, will be proved using the delay-integral-quadratic constraints in the following form. There exists a sequence $t_k \rightarrow +\infty$, possibly dependent on the functions $\sigma(\cdot)$ and $\xi(\cdot)$, such that

$$\int_0^{t_k} \mathcal{F}_j(\sigma(t),\xi(t),\sigma(t-\tau),\xi(t-\tau))dt + \gamma_j \geq 0, \forall \tau \in \mathbb{T}_j, j = 1,2,\ldots,N ,\tag{4.1.3}$$

except that, unlike Chap. 3, $\mathbb{T}_j$ is a countable subset of nonnegative real numbers.

D. Altshuller: Frequency Domain Criteria for Absolute Stability, LNCIS 432, pp. 81–115.
springerlink.com

Let $\vartheta_j(\tau)$ be a set of nondecreasing functions defined as follows. Let $\{\theta_{nj}\}$ for $j=1,2,\ldots N$ be a collection of N sequences $\{\theta_n\}$ with nonnegative terms. Set $\vartheta_j(\tau_0) = \theta_{0j}$; $\vartheta_j(\tau_{n+1}) = \vartheta_j(\tau_n) + \theta_{nj}$, and $\vartheta_j(\tau) = \vartheta_j(\tau_n)$ for $\tau \in (\tau_n; \tau_{n+1})$. Let the measures $\mu_j(\tau)$ in the frequency condition (2.8) be generated by these functions. Note that the measures thus defined satisfy the condition $\mu_j(\mathbb{R} \setminus \mathbb{T}_j) = 0$.

Then the integrals in (2.8) degenerate into series, and the frequency condition takes the following form. There exists a collection of sequences $\{\theta_{nj}\}$ such that for some constant $\varepsilon > 0$

$$\sum_{j=1}^{N} \sum_{n=1}^{\infty} \Pi_j(\omega, \tau_n)\theta_n + \varepsilon I \le 0, \tau_n \in \mathcal{T}. \tag{4.1.4}$$

For the sake of simplicity of notation, we shall write $\Pi_j(\omega, n)$ instead of $\Pi_j(\omega, \tau_n)$.

It is an immediate consequence of the theorems from Chap. 2 that if the system (4.1.1)-(4.1.2) is minimally stable and the frequency condition (4.1.4) is satisfied then this system is absolutely stable.

In this chapter we will take advantage of the periodicity of the nonlinearities to derive delay-integral-quadratic constraints, which will then be used to derive some stability criteria.

4.2 SISO Systems with Monotone Lipschitz Nonlinearities

In this section we consider SISO systems with nonlinearities satisfying the slope restriction condition:

$$0 \le \frac{\varphi(\sigma_1, t) - \varphi(\sigma_2, t)}{\sigma_1 - \sigma_2} \le \kappa, \varphi(0, t) \equiv 0. \tag{4.2.1}$$

We shall prove two results analogous to the ones proved in Chap. 3 for stationary systems. To this end, we need the following lemma.

Lemma 4.1. *Assume that the function* $\varphi(\sigma, t)$ *is periodic in t with the period T and satisfies the slope restriction condition (4.2.1). Denote* $\eta(t) \triangleq \kappa\sigma(t) - \varphi(\sigma(t), t)$

and $\xi(t) = \varphi(\sigma(t), t)$. *Then for any measurable function* $\sigma(t)$, *the following in-equalities hold for every number* b:

$$\int_0^b [\eta(t) - \eta(t-\tau)] \varphi(\sigma(t), t)) dt \geq 0;$$

$$\int_0^b [\xi(t) - \xi(t-\tau)] \eta(t) dt \geq 0.$$

If, in addition, the function $\varphi(\sigma)$ *is odd then the following inequalities holds for every number* b:

$$\int_0^b [\eta(t) + \eta(t-\tau)] \varphi(\sigma(t), t)) dt \geq 0;$$

$$\int_0^b [\xi(t) + \xi(t-\tau)] \eta(t) dt \geq 0.$$

Proof. Note that the functions $\varphi(\sigma, t)$ and $\kappa\sigma - \varphi(\sigma, t)$ satisfy automatically the conditions 1)-4) of Lemma 2.10. Since they are nondecreasing, they satisfy the condition 5) of Lemma 2.10 with $H(u, v) \equiv 0$. Application of Lemma 2.10 yields the desired conclusions.

Our objective will be first to prove two stability criteria for this type of systems and then to illustrate their geometric interpretation with some numerical examples. We follow, in part, the presentation from [11], which, in turn, offers some improvements over the results from [4].

4.2.1 Stability Multipliers

Let us proceed with the formulation and proof of the main results.

Theorem 4.2. *Assume the following:*

1) *The linear block* (4.1.1) *satisfies the regularity conditions;*
2) *The function* $\varphi(\sigma, t)$ *is continuous in each argument, satisfies condition* (4.2.1), *and* $\varphi(\sigma, t+T) = \varphi(\sigma, t)$;

3) There exists a series $\sum_{n=-\infty}^{n=+\infty} \theta_n < 1$ *with nonnegative terms such that for all real values of* ω

$$\text{Re}\left\{ \left[\kappa^{-1} + W(i\omega) \right] \left[1 - \sum_{n=-\infty}^{+\infty} \theta_n e^{-i\omega n T} \right] \right\} \geq \varepsilon > 0 . \tag{4.2.2}$$

Then for all functions $\sigma(\cdot)$ *and* $\xi(\cdot)$, *satisfying both* (4.1.1) *and* (4.1.2), $\sigma(\cdot) \in L^2(0;+\infty)$ *and, furthermore, there exists a positive constant* λ, *independent of the function* $\alpha(\cdot)$, *such that* $\|\sigma(\cdot)\| \leq \lambda \|\alpha(\cdot)\|$.

Proof. Define quadratic forms:

$$\mathcal{F}_1(\sigma_1, \xi_1) = \xi_1(\sigma_1 - \kappa^{-1}\xi_1);$$
$$\mathcal{F}_2(\sigma_1, \xi_1, \sigma_2, \xi_2) = (\xi_1 - \xi_2)(\sigma_1 - \kappa^{-1}\xi_1);$$
$$\mathcal{F}_3(\sigma_1, \xi_1, \sigma_2, \xi_2) = \left(\sigma_1 - \kappa^{-1}\xi_1 - \sigma_2 + \kappa^{-1}\xi_2 \right)\xi_1.$$

From the condition (4.2.1) it follows that

$$\mathcal{F}_1\big(\sigma(t), \xi(t)\big) \geq 0 . \tag{4.2.3}$$

In addition, Lemma 4.1 implies that the following inequality holds for $j = 2, 3$, any $t_k > 0$, and any $\tau = nT$:

$$\int_0^{t_k} \mathcal{F}_j\big(\sigma(t), \xi(t), \sigma(t-\tau), \xi(t-\tau)\big)\, dt \geq 0 . \tag{4.2.4}$$

Compute the matrices $\Pi_j(\omega, n)$ to obtain

$$\Pi_1(\omega, n) = -\text{Re}\left[\kappa^{-1} + W(i\omega) \right];$$
$$\Pi_2(\omega, n) = -\text{Re}\left\{ \left[\kappa^{-1} + W(i\omega) \right]\left(1 - e^{i\omega n T} \right) \right\};$$
$$\Pi_3(\omega, n) = -\text{Re}\left\{ \left[\kappa^{-1} + W(i\omega) \right]\left(1 - e^{-i\omega n T} \right) \right\}.$$

Therefore, the frequency condition takes the following form. There exist sequences θ_{n1}, θ_{n2}, and θ_{n3}, all with nonnegative terms, such that for all real values of ω

$$\mathrm{Re}\left\{\left[\kappa^{-1}+W(i\omega)\right]\left[\Theta-\sum_{n=0}^{+\infty}\theta_{n2}e^{i\omega nT}-\sum_{n=0}^{+\infty}\theta_{n2}e^{-i\omega nT}\right]\right\}\geq\varepsilon>0,$$

where $\Theta=\sum_{n=0}^{+\infty}\left(\theta_{n1}+\theta_{n2}+\theta_{n3}\right)$.

In order to see that this frequency condition is equivalent to the condition 3) of the theorem, set $\theta_n=\theta_{n2}/\Theta$ for $n>0$ and $\theta_n=\theta_{n3}/\Theta$ for $n<0$.

Minimal stability is verified in the same way as in Chap. 3 by constructing a bounded continuation for every process. Let $t_k\to+\infty$ be an arbitrary sequence. Define $m_k=\max_{t\in[0;t_k]}\left|\sigma(t)\right|$ and

$$\varphi_k(\sigma,t)=\begin{cases}\varphi(\sigma,t)\text{ if }\left|\sigma\right|\leq m_k\\ \varphi(m_k,t)\text{ if }\left|\sigma\right|>m_k\\ \varphi(-m_k,t)\text{ if }\left|\sigma\right|<-m_k\end{cases}.$$

The inequalities (4.2.3) and (4.2.4) hold for the process $z_k(\cdot)$ defined as a solution of the system consisting of (4.1.1) and

$$\xi(t)=\varphi_k(\sigma(t),t).$$

Therefore, the process $z_k(\cdot)$ is a bounded continuation of the process $z(\cdot)$. By Theorem 2.9 it is a stable continuation and since this construction is applicable to any process, the system is minimally stable.

By Theorem 2.6 the system (4.1.1)-(4.1.2) is absolutely stable, which is equivalent to the conclusion of the theorem. The proof is complete.

Similarly to the results in Chap. 3, if the function $\varphi(\sigma,t)$ is odd in σ, it is possible to relax the requirement that the terms of the sequence $\{\theta_n\}$ must be nonnegative.

Theorem 4.3. *Assume that hypotheses 1) and 2) of Theorem 4.2 hold and, in addition, the function $\varphi(\sigma,t)$ is odd in σ. Suppose further that there exists an absolutely convergent series $\sum_{n=-\infty}^{n=+\infty}\theta_n<1$ such that for all real values of ω the condition (4.2.2) holds.*

Then for all functions $\sigma(\cdot)$ and $\xi(\cdot)$, satisfying both (4.1.1) and (4.1.2), $\sigma(\cdot)\in L^2(0;+\infty)$ and, furthermore, there exists a positive constant λ, independent of the function $\alpha(\cdot)$, such that $\left\|\sigma(\cdot)\right\|\leq\lambda\left\|\alpha(\cdot)\right\|$.

Proof. First, we can define the same quadratic forms $\mathcal{F}_j(\sigma_1,\xi_1,\sigma_2,\xi_2)$, $j=1,2,3$ as in the proof of Theorem 4.2. The inequalities (4.2.3) and (4.2.4) still hold.

Define two additional quadratic forms:

$$\mathcal{F}_4\left(\sigma_1,\xi_1,\sigma_2,\xi_2\right)=\left(\xi_1+\xi_2\right)\left(\sigma_1-\kappa^{-1}\xi_1\right);$$
$$\mathcal{F}_5\left(\sigma_1,\xi_1,\sigma_2,\xi_2\right)=\left(\sigma_1-\kappa^{-1}\xi_1+\sigma_2-\kappa^{-1}\xi_2\right)\xi_1.$$

Using Lemma 4.1 we conclude that (4.2.4) holds for $j=4,5$ for any $\tau=nT$.
Computation of the matrices $\Pi_j(\omega,n)$ for $j=4,5$ yields

$$\Pi_4(\omega,n)=-\mathrm{Re}\left\{\left[\kappa^{-1}+W(i\omega)\right]\left(1+e^{i\omega nT}\right)\right\};$$
$$\Pi_5(\omega,n)=-\mathrm{Re}\left\{\left[\kappa^{-1}+W(i\omega)\right]\left(1+e^{-i\omega nT}\right)\right\}.$$

Therefore, the frequency condition takes the following form. There exist
sequences θ_{nj}, $j=1\ldots5$, all with nonnegative terms, such that for all real values
of ω

$$\mathrm{Re}\left\{\left[\kappa^{-1}+W(i\omega)\right]Z(i\omega)\right\}\geq\varepsilon>0.$$

Here

$$Z(i\omega)=\Theta-\sum_{n=0}^{+\infty}\theta_{n2}e^{i\omega nT}-\sum_{n=0}^{+\infty}\theta_{n3}e^{-i\omega nT}+\sum_{n=0}^{+\infty}\theta_{n4}e^{i\omega nT}+\sum_{n=0}^{+\infty}\theta_{n5}e^{-i\omega nT},$$

$$\Theta=\sum_{n=0}^{+\infty}\sum_{j=1}^{5}\theta_{nj}.$$

To show that this frequency condition is equivalent to the one of the theorem, set
$\theta_n=(\theta_{n2}-\theta_{n4})/\Theta$ for $n>0$ and $\theta_n=(\theta_{n3}-\theta_{n5})/\Theta$ for $n<0$. The proof is concluded in the same way as that of Theorem 4.2.

4.2.2 Lipatov Plots and Examples

As in Chap. 3, we can rewrite the frequency condition (4.2.2) in Theorems 4.2 and
4.3 in the form:

$$\left[\kappa^{-1}+\mathrm{Re}\,W(i\omega)\right]\mathrm{Re}\,Z(i\omega)-\mathrm{Im}\,W(i\omega)\,\mathrm{Im}\,Z(i\omega)>0,$$

where

$$Z(i\omega) = 1 - \sum_{n=-\infty}^{+\infty} \theta_n e^{-i\omega nT}$$

and the sequence $\{\theta_n\}$ satisfies the condition of either Theorem 4.2 or Theorem 4.3, depending on the context.

Also, similarly to Chap. 3, define two functions:

$$\Phi(\omega) = \frac{\kappa^{-1} + \operatorname{Re}W(i\omega)}{\operatorname{Im}W(i\omega)}, \Psi(\omega) = \frac{\operatorname{Im}Z(i\omega)}{\operatorname{Re}Z(i\omega)}.$$

The function $\Phi(\omega)$ is the same as in Chap. 3. However, the function $\Psi(\omega)$ now involves Fourier series instead of Fourier transforms:

$$\Psi(\omega) = \frac{\sum_{n=-\infty}^{+\infty} \theta_n \sin \omega nT}{\sum_{n=-\infty}^{+\infty} \theta_n \cos \omega nT - 1}. \tag{4.2.5}$$

Furthermore, $\sum_{n=-\infty}^{n=+\infty} \theta_n < 1$ and the coefficients θ_n must be nonnegative unless the function $\varphi(\sigma,t)$ is odd in σ.

It is certainly possible, and often easier, to set some of the coefficients θ_n in (4.2.5) to zero and use finite trigonometric sums instead of Fourier series.

Consider the system with the linear block described by the transfer function

$$W(s) = \frac{s^2}{\left[(s+0.5)^2 + 0.81\right]\left[(s+0.5)^2 + 1.21\right]}.$$

Let $\kappa = 11$. Choose $\theta_1 = 0.1, \theta_2 = 0.5, \theta_3 = 0.2$. Figure 4.1 shows the Lipatov plot for $T = 0.25\pi$. Since the curves do not intersect, the system is absolutely stable. The plot for $T = 0.3\pi$ is shown in Fig. 4.2, and, again, the conclusion is that the system is absolutely stable. The same is true for all the intermediate values of the period.

Consider now the case of an odd nonlinearity. This means that the coefficients θ_n can be negative. Choose $\theta_1 = -0.25, \theta_2 = 0.5, \theta_3 = 0.2$. The Lipatov plot for $T = 0.21\pi$ is shown in Fig. 4.3 and for $T = 0.3\pi$ - in Fig. 4.4. In both of these cases the system is absolutely stable as well as for all the intermediate values of the period. This illustrates how the introduction of the requirement that the nonlinearity is odd leads to widening of the range of values of the period for which the system is absolutely stable.

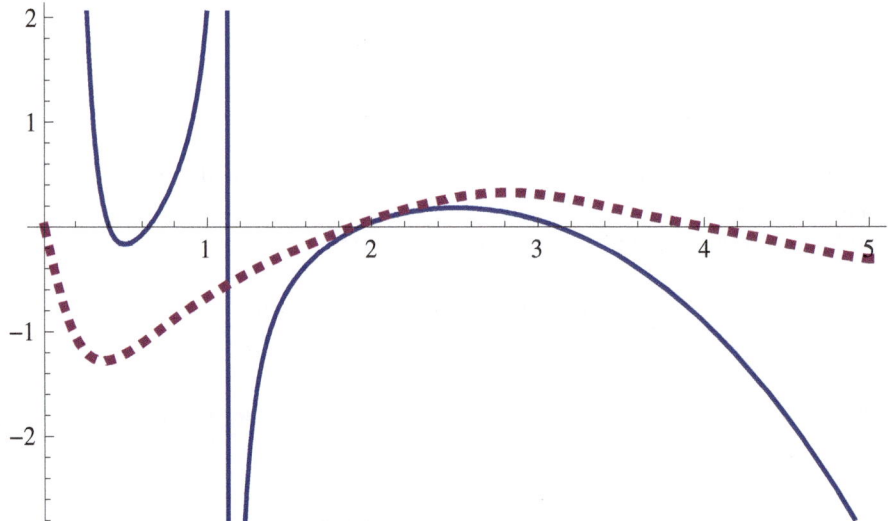

Fig. 4.1 Lipatov plot for $T=0.25\pi$. Function $\Phi(\omega)$ is shown as a solid line and function $\Psi(\omega)$ is shown as a broken line

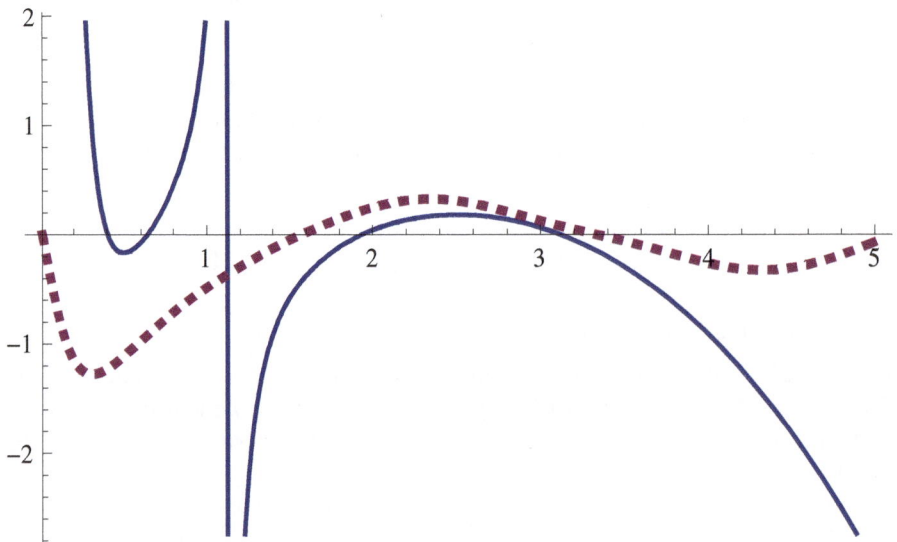

Fig. 4.2 Lipatov plot for $T=0.3\pi$. Function $\Phi(\omega)$ is shown as a solid line and function $\Psi(\omega)$ is shown as a broken line

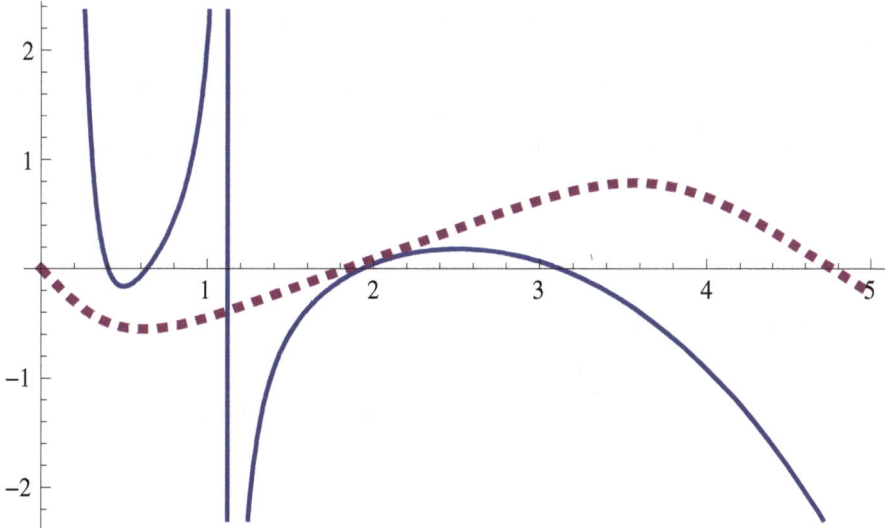

Fig. 4.3 Lipatov plot for odd nonlinearity and $T=0.21\pi$. Function $\Phi(\omega)$ is shown as a solid line and function $\Psi(\omega)$ is shown as a broken line

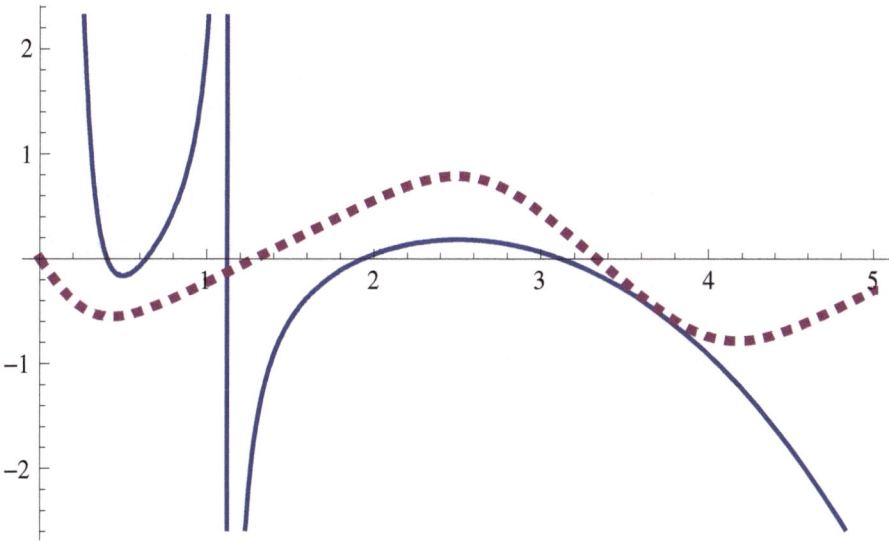

Fig. 4.4 Lipatov plot for odd nonlinearity and $T=0.3\pi$. Function $\Phi(\omega)$ is shown as a solid line and function $\Psi(\omega)$ is shown as a broken line

4.3 MIMO Systems with Gradient Nonlinearities

Stability criteria proved in the previous section can be extended to MIMO systems with gradient nonlinearities. Two results have the form, similar to Theorems 3.16 and 3.17 and are proved in essentially the same way. The slope restriction condition (4.2.1) is replaced with

$$0 \le \frac{\partial \varphi(\sigma,t)}{\partial \sigma} < \Lambda \,.$$ (4.3.1)

Theorem 4.4. *Assume the following:*

1) The linear block (4.1.1) satisfies the regularity conditions;
2) The function $\varphi(\sigma,t)$ is differentiable, satisfies the condition (4.3.1),
$\varphi(\sigma,t+T) \equiv \varphi(\sigma,t)$, *and*

$$\left[\frac{\partial \varphi(\sigma,t)}{\partial \sigma}\right]^{*} - \frac{\partial \varphi(\sigma,t)}{\partial \sigma} \equiv 0 \,;$$ (4.3.2)

3) There exists a series $\sum_{n=-\infty}^{n=+\infty} \theta_n < 1$ with nonnegative terms such that for all real values of ω

$$\text{Re}\left\{\left[1 - \sum_{n=-\infty}^{+\infty} \theta_n e^{-i\omega nT}\right]\left[\Lambda^{-1} + W(i\omega)\right]\right\} \ge \varepsilon > 0 \,.$$ (4.3.3)

Then for all functions $\sigma(\cdot)$ and $\xi(\cdot)$, satisfying both (4.1.1) and (4.1.2), $\sigma(\cdot) \in L^2(0;+\infty)$ and, furthermore, there exists a positive constant λ, independent of the function $\alpha(\cdot)$, such that $\|\sigma(\cdot)\| \le \lambda \|\alpha(\cdot)\|$.

Proof. First, define a quadratic form:

$$\mathcal{F}_1(\sigma_1,\xi_1) = \xi_1^*\left(\sigma_1 - \Lambda^{-1}\xi_1\right).$$

Condition (4.3.1) implies that

$$\mathcal{F}_1(\sigma(t),\xi(t)) \ge 0 \,.$$ (4.3.4)

Indeed, by Mean Value Theorem, we have for some vector $\bar{\sigma}(t)$

$$\xi(t) = \frac{\partial \varphi(\sigma,t)}{\partial \sigma}\bigg|_{\sigma=\bar{\sigma}(t)} \sigma(t) \,.$$

Because of (4.3.1), the matrix $\partial\varphi(\sigma,t)/\partial\sigma$ is positive-semidefinite and the matrix $I - \Lambda^{-1}[\partial\varphi(\sigma,t)/\partial\sigma]$ is positive-definite. Therefore,

$$\xi^*(t)[\sigma(t) - \Lambda^{-1}\xi(t)]$$

$$= \sigma^*(t)\left[\left.\frac{\partial\varphi(\sigma,t)}{\partial\sigma}\right|_{\sigma=\bar{\sigma}(t)}\right]^*\left[I - \Lambda^{-1}\left.\frac{\partial\varphi(\sigma,t)}{\partial\sigma}\right|_{\sigma=\bar{\sigma}(t)}\right]\sigma(t)$$

$$\geq 0.$$

Next, define the quadratic forms:

$$\mathcal{F}_2(\sigma_1,\xi_1,\sigma_2,\xi_2) = (\sigma_1 - \Lambda^{-1}\xi_1)^*(\xi_1 - \xi_2);$$
$$\mathcal{F}_3(\sigma_1,\xi_1,\sigma_2,\xi_2) = \xi_1^*(\sigma_1 - \Lambda^{-1}\xi_1 - \sigma_2 - \Lambda^{-1}\xi_2).$$

Let us prove that for any $t_k > 0$

$$\int_0^{t_k}\mathcal{F}_j\big(\sigma(t),\xi(t),\sigma(t-nT),\xi(t-nT)\big)dt \geq 0, \quad j = 2,3. \qquad (4.3.5)$$

Because of (4.3.2), we can define the function $\Phi(\sigma,t)$ as a path integral:

$$\Phi(\sigma,t) = \int_0^{\sigma}\varphi(\sigma,t)d\sigma. \qquad (4.3.6)$$

This function is convex since its Hessian is the Jacobian of the function $\varphi(\sigma,t)$, which is positive-semidefinite by (4.3.1).

We can use convexity of the function $\Phi(\sigma,t)$ and apply Lemma 2.10 by setting

$$x(t) = \varphi\big(\sigma(t),t\big);$$
$$g\big(x(t),t\big) = \sigma(t) - \Lambda^{-1}\varphi\big(\sigma(t),t\big)$$

for $j = 2$ and

$$x(t) = \sigma(t) - \Lambda^{-1}\varphi\big(\sigma(t),t\big);$$
$$g\big(x(t),t\big) = \varphi\big(\sigma(t),t\big)$$

for $j = 3$. It is easy to verify via Implicit Function Theorem that all the relevant functions are well defined and, furthermore, they satisfy the hypotheses of Lemma 2.10. This proves (4.3.5).

We can now compute the matrices $\Pi_j(\omega, n)$ to obtain

$$\Pi_1(\omega, n) = -\operatorname{Re}\left[\Lambda^{-1} + W(i\omega)\right];$$

$$\Pi_2(\omega, n) = -\operatorname{Re}\left\{\left(1 - e^{i\omega nT}\right)\left[\Lambda^{-1} + W(i\omega)\right]\right\};$$

$$\Pi_3(\omega, n) = -\operatorname{Re}\left\{\left(1 - e^{-i\omega nT}\right)\left[\Lambda^{-1} + W(i\omega)\right]\right\}.$$

The frequency condition now takes the following form. There exist sequences θ_{n1}, θ_{n2}, and θ_{n3}, all with nonnegative terms, such that for all real values of ω

$$\operatorname{Re}\left\{\left[\Theta - \sum_{n=0}^{+\infty}\theta_{n2}e^{i\omega nT} - \sum_{n=0}^{+\infty}\theta_{n2}e^{-i\omega nT}\right]\left[\Lambda^{-1} + W(i\omega)\right]\right\} \geq \varepsilon I > 0,$$

where $\Theta = \sum_{n=0}^{+\infty}\left(\theta_{n1} + \theta_{n2} + \theta_{n3}\right)$.

Just as before, set $\theta_n = \theta_{n2}/\Theta$ for $n>0$ and $\theta_n = \theta_{n3}/\Theta$ for $n<0$ to see that this condition is equivalent to condition 3) of the theorem. The final step of the proof is the verification of minimal stability, which is accomplished in exactly the same way as in the proof of Theorem 4.2.

Similarly to the earlier results, if the nonlinearity is odd, it is possible to weaken the requirement that the terms θ_n must be nonnegative.

Theorem 4.5. *Assume that the conditions 1) and 2) of Theorem 4.4 hold, and, in addition, $\varphi(-\sigma, t) \equiv \varphi(\sigma, t)$. Suppose further that there exists an absolutely convergent series $\sum_{n=-\infty}^{n=+\infty}\theta_n < 1$ such that for all real values of ω, the inequality (4.3.3) holds.*

Then for all functions $\sigma(\cdot)$ and $\xi(\cdot)$, satisfying both (4.1.1) and (4.1.2), $\sigma(\cdot) \in L^2(0; +\infty)$ and, furthermore, there exists a positive constant λ, independent of the function $\alpha(\cdot)$, such that $\|\sigma(\cdot)\| \leq \lambda \|\alpha(\cdot)\|$.

Proof. First, we define the same quadratic forms $\mathcal{F}_j(\sigma_1, \xi_1, \sigma_2, \xi_2)$, $j=1,2$, and 3 as in the proof of Theorem 4.4. The inequalities (4.3.4) and (4.3.5) apply.

Define two additional quadratic forms:

$$\mathcal{F}_4(\sigma_1, \xi_1, \sigma_2, \xi_2) = \left(\sigma_1 - \Lambda^{-1}\xi_1\right)^*\left(\xi_1 + \xi_2\right);$$

$$\mathcal{F}_5(\sigma_1, \xi_1, \sigma_2, \xi_2) = \xi_1^*\left(\sigma_1 - \Lambda^{-1}\xi_1 + \sigma_2 - \Lambda^{-1}\xi_2\right).$$

Once again, we can apply Lemma 2.10 to the function $\Phi(\sigma,t)$ defined by (4.3.6) with

$$x(t) = \varphi\big(\sigma(t),t\big);$$
$$g\big(x(t),t\big) = \sigma(t) - \Lambda^{-1}\varphi\big(\sigma(t),t\big)$$

for $j = 4$ and

$$x(t) = \sigma(t) - \Lambda^{-1}\varphi\big(\sigma(t),t\big);$$
$$g\big(x(t),t\big) = \varphi\big(\sigma(t),t\big)$$

for $j = 5$.

Computation of the matrices $\Pi_j(\omega,n)$ yields

$$\Pi_4(\omega,n) = -\operatorname{Re}\left\{\big(1+e^{i\omega nT}\big)\big[\Lambda^{-1}+W(i\omega)\big]\right\};$$
$$\Pi_5(\omega,n) = -\operatorname{Re}\left\{\big(1+e^{-i\omega nT}\big)\big[\Lambda^{-1}+W(i\omega)\big]\right\}.$$

Therefore, the frequency condition takes the following form. There exist sequences θ_{nj}, $j = 1\ldots 5$, all with nonnegative terms, such that for all real values of ω

$$\operatorname{Re}\left\{Z(i\omega)\big[\Lambda^{-1}+W(i\omega)\big]\right\} \geq \varepsilon > 0.$$

Here

$$Z(i\omega) = \Theta - \sum_{n=0}^{+\infty}\theta_{n2}e^{i\omega nT} - \sum_{n=0}^{+\infty}\theta_{n3}e^{-i\omega nT} + \sum_{n=0}^{+\infty}\theta_{n4}e^{i\omega nT} + \sum_{n=0}^{+\infty}\theta_{n5}e^{-i\omega nT},$$
$$\Theta = \sum_{n=0}^{+\infty}\sum_{j=1}^{5}\theta_{nj}.$$

To show that this frequency condition is equivalent to the one of the theorem, set $\theta_n = (\theta_{n2}-\theta_{n4})/\Theta$ for $n>0$ and $\theta_n = (\theta_{n3}-\theta_{n5})/\Theta$ for $n<0$. The proof is concluded in the same way as that of Theorem 4.4.

We may apply this result to derive a criterion for stability of periodic solutions of the dynamical system

$$\dot{x} = Ax + \phi(x) + f(t). \tag{4.3.7}$$

Here $x(t) \in \mathbb{R}^m$ and $f(t)$ is a periodic function with a period T (it may be identically equal to zero, but then there is a question of existence of periodic solutions, which we do not address here). Assume that the matrix A is Hurwitz stable and that $\phi(x)$ is differentiable and satisfies both conditions (4.3.1) and (4.3.2). Define the transfer matrix $W(s) = [sI - A]^{-1}$.

It is well known that (4.3.7) has a periodic solution with a period T, which we denote $x_0(t)$. Define the new variable $\sigma(t) = x(t) - x_0(t)$ and the function

$$\varphi(x,t) = \phi(\sigma + x_0(t)) - \phi(x_0(t)) .$$

This yields the following equation:

$$\dot{\sigma} = A\sigma + \varphi(\sigma,t) . \tag{4.3.8}$$

Applying Theorem 4.4 we conclude that the zero solution of (4.3.8) and, therefore, the periodic solution $x_0(t)$ of (4.3.7) is stable if the FC of Theorem 4.4 holds. Stability in this case is understood in the following sense. Let $x(t)$ be any other solution of (4.3.7). Then $|x(t) - x_0(t)| \in L^2(0;+\infty)$ and, furthermore, there exists a constant λ, the same for all solutions of (4.3.7), such that

$$|x(t) - x_0(t)| \in L^2(0;+\infty) .$$

It is worth noting that this stability condition does not require finding an explicit form of the solution $x_0(t)$, a usually impossible task. Therefore, it offers some advantage over the traditional Floquet multiplier method. This result is an extension of the one from the earlier paper [9]. This approach to derive stability criteria for forced oscillations was first used by Yakubovich in [152].

4.4 SISO Systems with Quasimonotone Nonlinearities

Thus far in the book, we have considered mostly monotone nonlinearities satisfying the Lipschitz condition, sometimes referred to as the slope restriction. Proofs of all stability criteria relied on Lemma 2.10 to establish that nonlinearity satisfied certain quadratic constraints. Recall that one of the conditions of this lemma is the inequality:

$$G(y,t) - G(x,y) + (\zeta x - y)g(x,t) \geq -H(x,g(x,t)) - H(y,g(y,t)) . \tag{4.4.1}$$

For monotone Lipschitz nonlinearities, this inequality was satisfied with $\zeta = 1$ and $H(u,v) \equiv 0$. In this section we consider systems with nonlinearities satisfying (4.4.1) without these two restrictions. We will call such functions quasimonotone.

The concept of a quasimonotone function was introduced by Barabanov [24]. He considered functions, satisfying (4.4.1) with $\zeta = 1$, but with the function $H(u,v)$ being a positive-semidefinite quadratic form. Some results for this type of nonlinearity were proved earlier in [5] with improvements given in [11]. A particular case of functions satisfying (4.4.1) with $H(u,v) \equiv 0$, but with $\zeta > 1$ was first considered by Rantzer [123] and further considered by Kulkarni et al [74]. Materassi and Salapaka [97] investigated the case with $\zeta < 1$.

In this section we will investigate stability of systems with time-periodic nonlinearities satisfying both of these two types of quasimonotonicity. Throughout this section it will be assumed that the nonlinear block satisfies the sector condition

$$0 \leq \frac{\varphi(\sigma,t)}{\sigma} \leq \kappa. \tag{4.4.2}$$

4.4.1 Quasimonotone Nonlinearities in the Sense of Barabanov

Paraphrasing from the paper by Barabanov [24], let us introduce the following definition.

Definition 4.6. *A function g(x) is called quasimonotone in the sense of Barabanov (hereafter, simply quasimonotone) with a defining form $\mathcal{G}(u,v)$ if there exists a positive-semidefinite quadratic form $\mathcal{G}(u,v)$ such that for all x and y*

$$G(y) - G(x) + (x-y)g(x) \geq -\mathcal{G}\big(x, g(x)\big) - \mathcal{G}\big(y, g(y)\big), \tag{4.4.3}$$

where

$$G(x) = \int_0^x g(x)dx.$$

Clearly, if $\mathcal{G}(u,v) \equiv 0$, the condition (4.4.3) reduces to the condition that the function $G(x)$ is concave, which implies that the function $g(x)$ is nondecreasing.

The main result of this subsection is stated as follows.

Theorem 4.7. *Assume the following:*

1) The linear block (4.1.1) satisfies the regularity conditions;
2) The function $\varphi(\sigma,t)$ is continuous in each argument, is quasimonotone in σ in the sense of Barabanov with a defining form $\mathcal{G}(r,w)$, is periodic in t with a period T, and satisfies the sector condition (4.4.2);

3) There exists a series $\sum_{n=-\infty}^{+\infty} \theta_n < 1$ *with nonnegative terms such that for all real values of* ω

$$\mathrm{Re}\left\{ \left[\zeta^{-1} + W(i\omega) \right]\left[1 - \sum_{n=-\infty}^{+\infty} \theta_n e^{-i\omega nT} \right] - 2\mathcal{G}\left(W(i\omega),1\right) \right\} \geq \varepsilon > 0. \quad (4.4.4)$$

Then for all functions $\sigma(\cdot)$ *and* $\xi(\cdot)$, *satisfying both* (4.1.1) *and* (4.1.2), $\sigma(\cdot) \in L^2(0;+\infty)$ *and, furthermore, there exists a positive constant* λ, *independent of the function* $\alpha(\cdot)$, *such that* $\|\sigma(\cdot)\| \leq \lambda \|\alpha(\cdot)\|$.

Proof. First, define the quadratic form:

$$\mathcal{F}_1\left(\sigma_1,\xi_1\right) = \xi_1\left(\sigma_1 - \kappa^{-1}\xi_1\right).$$

Sector condition (4.4.2) implies that

$$\mathcal{F}_1\left(\sigma_1(t),\xi_1(t)\right) \geq 0. \quad (4.4.5)$$

Next, define two more quadratic forms:

$$\mathcal{F}_2\left(\sigma_1,\xi_1,\sigma_2,\xi_2\right) = \left(\xi_1 - \xi_2\right)\left(\sigma_1 - \kappa^{-1}\xi_1\right) + \mathcal{G}\left(\sigma_1,\xi_1\right) + \mathcal{G}\left(\sigma_2,\xi_2\right);$$
$$\mathcal{F}_3\left(\sigma_1,\xi_1,\sigma_2,\xi_2\right) = \left(\sigma_1 - \kappa^{-1}\xi_1 - \sigma_2 + \kappa^{-1}\xi_2\right)\xi_1 + \mathcal{G}\left(\sigma_1,\xi_1\right) + \mathcal{G}\left(\sigma_2,\xi_2\right).$$

Using Lemma 2.10 with $H(u,v) = \mathcal{G}(u,v)$, we find that for any t_k

$$\int_0^{t_k} \mathcal{F}_j\left(\sigma(t),\xi(t),\sigma(t-\tau),\xi(t-\tau)\right)dt \geq 0 , \, j=2,3. \quad (4.4.6)$$

Compute the matrices $\Pi_j(\omega,n)$ to obtain

$$\Pi_1(\omega,n) = -\mathrm{Re}\left[\kappa^{-1} + W(i\omega) \right];$$
$$\Pi_2(\omega,n) = -\mathrm{Re}\left\{ \left[\kappa^{-1} + W(i\omega) \right]\left(1 - e^{i\omega nT}\right) - 2\mathcal{G}\left(W(i\omega),1\right) \right\};$$
$$\Pi_3(\omega,n) = -\mathrm{Re}\left\{ \left[\kappa^{-1} + W(i\omega) \right]\left(1 - e^{-i\omega nT}\right) - 2\mathcal{G}\left(W(i\omega),1\right) \right\}.$$

Therefore, the frequency condition takes the following form. There exist sequences θ_{n1}, θ_{n2}, and θ_{n3}, all with nonnegative terms, such that for all real values of ω

$$\text{Re}\left\{\left[\kappa^{-1}+W(i\omega)\right]\left[\Theta-\sum_{n=0}^{+\infty}\theta_{n2}e^{i\omega nT}-\sum_{n=0}^{+\infty}\theta_{n3}e^{-i\omega nT}\right]-2\mathcal{G}\left(W(i\omega),1\right)\right\}\geq\varepsilon>0,$$

where $\Theta=\sum_{n=0}^{+\infty}\left(\theta_{n1}+\theta_{n2}+\theta_{n3}\right)$.

In order to see that this frequency condition is equivalent to the condition 3) of the theorem, set $\theta_n=\theta_{n2}/\Theta$ for $n>0$ and $\theta_n=\theta_{n3}/\Theta$ for $n<0$.

In order to verify minimal stability, we construct a bounded continuation for every process as follows. Let $t_k\to\infty$ be an arbitrary sequence and set $m_k=\max_{t\in[0;t_k]}|\sigma(t)|$. Let s_k be a real number, such that for all t and all $\sigma\in[0;m_k]$

$$\left|\varphi(s_k,t)\right|\geq\left|\varphi(\sigma,t)\right|.$$

Define the functions

$$\varphi_k(\sigma,t)=\begin{cases}\varphi(-s_k,t) & \text{if}\quad \sigma<-m_k \\ \varphi(\sigma,t) & \text{if}\quad |\sigma|\leq m_k \\ \varphi(s_k,t) & \text{if}\quad \sigma>m_k.\end{cases}$$

These functions may have discontinuities of the first kind in the variable σ while the nonlinearity $\varphi(\sigma,t)$ was assumed to be continuous in each argument. However, this is not a concern since the functions $\varphi_k(\sigma,t)$ can be approximated by continuous functions with a level of precision sufficient for the relevant inequalities to hold.

Now we have to prove that the functions $\varphi_k(\sigma,t)$ are quasimonotone in σ with the same defining form $\mathcal{G}(r,w)$ as the function $\varphi(\sigma,t)$. This means proving that for all u and v the following inequality holds:

$$\int_u^v\varphi_k(\sigma,t)d\sigma+(u-v)\varphi_k(u,t)\geq-\mathcal{G}(u,\varphi_k(u,t))-\mathcal{G}(v,\varphi_k(v,t)).\quad(4.4.7)$$

We shall give a detailed proof for the case when $0<u<v$. Proofs in other cases are similar. If $u<v\leq m_k$, then $\varphi_k(\sigma,t)=\varphi(\sigma,t)$ on the entire interval of integration,

and (4.4.7) holds by assumption of the theorem. Another simple case is $m_k \le u < v$, in which the left-hand side of (4.4.7) vanishes, while the right-hand side is nonpositive.

Now let $u < m_k < v$. We have

$$\int_u^v \varphi_k(\sigma,t)\,d\sigma + (u-v)\varphi_k(u,t)$$

$$= \int_u^{m_k} \varphi_k(\sigma,t)\,d\sigma + \int_{m_k}^v \varphi_k(\sigma,t)\,d\sigma + (u-m_k)\varphi_k(u,t) + (m_k-v)\varphi_k(u,t).$$

On the interval $(u; m_k)$, the functions $\varphi_k(\sigma,t)$ and $\varphi(\sigma,t)$ coincide. Therefore, by definition of a quasimonotone function,

$$\int_u^{m_k} \varphi_k(\sigma,t)\,d\sigma + (u-m_k)\varphi_k(u,t) \ge -\mathcal{G}(u,\varphi_k(u,t)) - \mathcal{G}(m_k,\varphi_k(m_k,t)).$$

On the interval $(m_k; v)$, the function $\varphi_k(\sigma,t)$ is equal to $\varphi(s_k,t)$. Hence

$$\int_{m_k}^v \varphi_k(\sigma,t)\,d\sigma + (m_k-v)\varphi_k(u,t) = (v-m_k)\big[\varphi(s_k,t) - \varphi(u,t)\big] \ge 0.$$

Note that $\varphi_k(v,t) = \varphi(s_k,t) \ge \varphi(m_k,t) = \varphi_k(m_k,t)$. Therefore, since v and m_k are assumed to be positive, $\mathcal{G}(m_k,\varphi_k(m_k,t)) \le \mathcal{G}(v,\varphi_k(v,t))$. Putting it all together, we conclude that (4.4.7) is true in this case as well.

Having proved that the functions $\varphi_k(\sigma,t)$ are quasimonotone in σ with the same defining form $\mathcal{G}(r,w)$ as the function $\varphi(\sigma,t)$, we have thus exhibited a bounded continuation for an arbitrary process, which establishes that the system is minimally stable. By Theorem 2.6 the system (4.1.1)-(4.1.2) is absolutely stable, which is equivalent to the conclusion of the theorem. The proof is complete.

Similarly to the results of the previous section, if the nonlinearity is odd in σ, we can weaken the requirement that the coefficients θ_n must be nonnegative.

Theorem 4.8. *Suppose that the conditions 1) and 2) of Theorem 4.7 are met, and, in addition, the function $\varphi(\sigma,t)$ is odd in σ. Assume further that there exists an absolutely convergent series $\sum_{n=-\infty}^{+\infty} \theta_n < 1$ such that for all real values of ω*

$$\mathrm{Re}\left\{\left[\kappa^{-1} + W(i\omega)\right]\left[1 - \sum_{n=-\infty}^{+\infty} \theta_n e^{-i\omega n T}\right] - 2\mathcal{G}(W(i\omega),1)\right\} \ge \varepsilon > 0. \quad (4.4.8)$$

Then for all functions $\sigma(\cdot)$ *and* $\xi(\cdot)$, *satisfying both* (4.1.1) *and* (4.1.2), $\sigma(\cdot) \in L^2(0;+\infty)$ *and, furthermore, there exists a positive constant* λ, *independent of the function* $\alpha(\cdot)$, *such that* $\|\sigma(\cdot)\| \le \lambda \|\alpha(\cdot)\|$.

Proof. The proof of this theorem proceeds along the same step as the previous one. First, we define the same quadratic forms $\mathcal{F}_1$, $\mathcal{F}_2$, and $\mathcal{F}_3$. The inequalities (4.4.5) and (4.4.6) hold.

Define two more quadratic forms:

$$\mathcal{F}_4\left(\sigma_1,\xi_1,\sigma_2,\xi_2\right) = \left(\xi_1+\xi_2\right)\left(\sigma_1-\kappa^{-1}\xi_1\right)+\mathcal{G}\left(\sigma_1,\xi_1\right)+\mathcal{G}\left(\sigma_2,\xi_2\right);$$
$$\mathcal{F}_5\left(\sigma_1,\xi_1,\sigma_2,\xi_2\right) = \left(\sigma_1-\kappa^{-1}\xi_1+\sigma_2-\kappa^{-1}\xi_2\right)\xi_1+\mathcal{G}\left(\sigma_1,\xi_1\right)+\mathcal{G}\left(\sigma_2,\xi_2\right).$$

Since the function $\varphi(\sigma,t)$ is odd in σ, Lemma 2.10 implies that (4.4.6) holds for $j = 4,5$. Compute the matrices $\Pi_j(\omega,n)$ to obtain

$$\Pi_4(\omega,n) = -\mathrm{Re}\left\{\left[\kappa^{-1}+W(i\omega)\right]\left(1+e^{i\omega nT}\right)-2\mathcal{G}\left(W(i\omega),1\right)\right\};$$
$$\Pi_5(\omega,n) = -\mathrm{Re}\left\{\left[\kappa^{-1}+W(i\omega)\right]\left(1+e^{-i\omega nT}\right)-2\mathcal{G}\left(W(i\omega),1\right)\right\}.$$

Therefore the frequency condition takes the following form. There exist sequences θ_{n1}, θ_{n2}, and θ_{n3}, all with nonnegative terms, such that for all real values of ω

$$\mathrm{Re}\left\{\left[\kappa^{-1}+W(i\omega)\right]Z(i\omega)-2\mathcal{G}\left(W(i\omega),1\right)\right\} \ge \varepsilon > 0,$$

where

$$Z(i\omega) = \Theta - \sum_{n=0}^{+\infty}\theta_{n2}e^{i\omega nT} - \sum_{n=0}^{+\infty}\theta_{n3}e^{-i\omega nT} + \sum_{n=0}^{+\infty}\theta_{n4}e^{i\omega nT} + \sum_{n=0}^{+\infty}\theta_{n5}e^{-i\omega nT}$$

and

$$\Theta = \sum_{n=0}^{+\infty}\left(\theta_{n1}+\theta_{n2}+\theta_{n3}+\theta_{n4}+\theta_{n5}\right).$$

Set $\theta_n = (\theta_{2n}-\theta_{4n})/\Theta$ for $n>0$ and $\theta_n = (\theta_{3n}-\theta_{5n})/\Theta$ for $n<0$ to obtain the frequency condition of the theorem.

To verify minimal stability, we use the same functions $\varphi_k(\sigma,t)$ as in the proof of Theorem 4.5. They are odd in σ if the function $\varphi(\sigma,t)$ is odd in σ. Therefore, the proof can be completed in the same way as the proof of Theorem 4.7.

Barabanov [24] proved similar stability criteria for stationary systems. Theorems 4.7 and 4.8 extend his results to the case of time-periodic nonlinearities. Note that the stability criteria in these two theorems do not have the multiplier form.

4.4.2 The Second Type of Quasimonotone Nonlinearities

In this subsection we consider the other type of quasimonotone nonlinearities, i.e., the case when (4.4.1) is satisfied with $\zeta \neq 1$. An example of one such function, considered by Rantzer [123], is shown in Fig. 4.5 (left). In this case $\zeta = 1 + \delta$. Kulkarni et al. [74] showed that this function satisfies (4.4.1) with $\zeta = 1 + 2\delta$.

Another type of nonlinearities satisfying this condition was introduced by Materassi and Salapaka [97]. They imposed the following conditions. Two additional functions $\overline{\varphi}(\sigma, t)$ and $\delta(\sigma)$ are introduced, such that

1) $\varphi(\sigma, t) = \overline{\varphi}(\sigma, t)\big[1 + \delta(\sigma)\big]$;

2) $\overline{\varphi}(\sigma, t)$ is nondecreasing and odd in σ;

3) $\forall \sigma \in \mathbb{R}: \ \big|\delta(\sigma)\big| \leq D < 1$.

The constant D is called the spread.

An example of the function satisfying these conditions is

$$\varphi(\sigma, t) = \arctan(\sigma)\big[1 + D\sin(\alpha\sigma + \phi)\big] f(t),$$

where $\overline{\varphi}(\sigma, t) = \arctan(\sigma)$, $\delta(\sigma) = D\sin(\alpha\sigma + \phi)$, $D \geq 0$, and $-\pi \leq \phi \leq \pi$. It is shown in Fig. 4.5 (right). This figure also shows with dashed lines the plots of functions $[1 \pm D]\overline{\varphi}(\sigma, t)$.

Nonlinearities of this type occur when interference fringes disturb measurements obtained by photodiodes [1, 96].

Careful perusal of the proofs in [97] makes it apparent that the functions satisfying these conditions also satisfy (4.4.1) with

$$\zeta = \frac{1 - D}{1 + D}.$$

It is worth noting that in Rantzer's example, $\zeta > 1$ while $\zeta < 1$ in [97]. The intermediate case of $\zeta = 1$ corresponds to monotone Lipschitz nonlinearities.

Define

$$\Phi(\sigma, t) = \int_0^\sigma \varphi(\sigma, t) d\sigma.$$

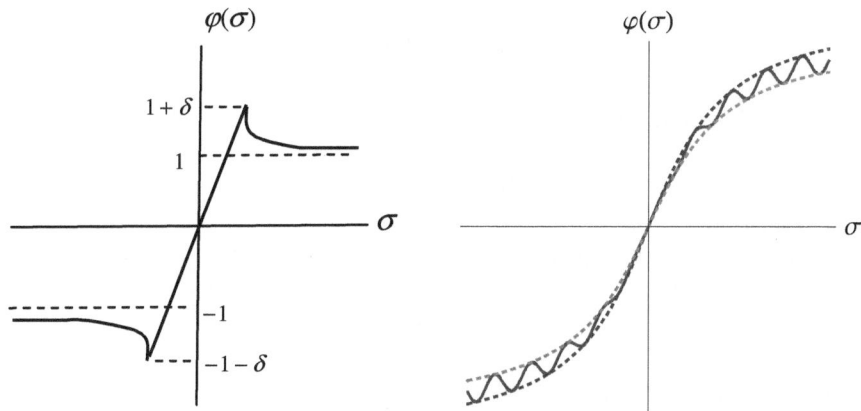

Fig. 4.5 Examples of quasimonotone functions

For systems with nonlinearities of this type, we have the following stability criterion.

Theorem 4.9. *Assume the following:*

1) The linear block (4.1.1) satisfies the regularity conditions;
2) The function $\varphi(\sigma,t)$ is continuous in each argument, is periodic in t with a period T, and satisfies the sector condition (4.4.2):
3) There exists a number $\zeta > 0$, such that for all values of σ_1, σ_2, and t:

$$\zeta\sigma_1\varphi(\sigma_1,t) - \sigma_2\varphi(\sigma_1,t) \geq \Phi(\sigma_1,t) - \Phi(\sigma_2,t)$$

4) There exists a series $\sum_{n=0}^{+\infty}\theta_n < 1/\zeta$ with nonnegative terms, such that for all real values of ω

$$\kappa^{-1} + \mathrm{Re}\left\{W(i\omega)\left[1 - \sum_{n=0}^{+\infty}\theta_n e^{i\omega nT}\right]\right\} \geq \varepsilon > 0. \qquad (4.4.9)$$

Then for all functions $\sigma(\cdot)$ and $\xi(\cdot)$, satisfying both (4.1.1) and (4.1.2), $\sigma(\cdot) \in L^2(0;+\infty)$ and, furthermore, there exists a positive constant λ, independent of the function $\alpha(\cdot)$, such that $\|\sigma(\cdot)\| \leq \lambda\|\alpha(\cdot)\|$.

Proof. As before, define the quadratic form:

$$\mathcal{F}_1\left(\sigma_1,\xi_1\right)=\xi_1\left(\sigma_1-\kappa^{-1}\xi_1\right).$$

Sector condition (4.4.2) implies that

$$\mathcal{F}_1\left(\sigma_1(t),\xi_1(t)\right)\ge 0. \qquad (4.4.10)$$

Next, define the following quadratic form:

$$\mathcal{F}_2(\sigma_1,\xi_1,\sigma_2,\xi_2)=\zeta\sigma_1\xi_1-\sigma_2\xi_1.$$

Lemma 2.10 implies that for all $t>0$

$$\int_0^t \mathcal{F}_2(\sigma(t),\xi(t),\sigma(t-T),\xi(t-T)dt \ge 0. \qquad (4.4.11)$$

Computation of the matrices $\Pi_1(\omega,n)$ and $\Pi_2(\omega,n)$ yields

$$\Pi_1(\omega,n)=-\operatorname{Re}\left[\kappa^{-1}+W(i\omega)\right];$$
$$\Pi_2(\omega,n)=-\operatorname{Re}\left[\left(\zeta-e^{-i\omega nT}\right)W(i\omega)\right].$$

Therefore, the frequency condition takes the following form. There exist sequences θ_{n1} and θ_{n2}, both with nonnegative terms, such that for all real values of ω

$$\operatorname{Re}\left[\kappa^{-1}+W(i\omega)\right]\sum_{n=0}^{\infty}\theta_{1n}+\sum_{n=0}^{\infty}\theta_{2n}\operatorname{Re}\left[\left(\zeta-e^{-i\omega nT}\right)W(i\omega)\right]\ge\varepsilon>0.$$

Set, without loss of generality, $\kappa^{-1}\sum_{n=0}^{\infty}\theta_{1n}+\zeta\sum_{n=0}^{\infty}\theta_{2n}=\kappa^{-1}$ and $\theta_n=\theta_{2n}$ to see that this frequency condition is equivalent to (4.4.9).

The final step in proving this theorem is to construct a bounded continuation for an arbitrary process $z(\bullet)=[\sigma(\bullet),\xi(\bullet)]$ such that $\xi(t)=\varphi(\sigma(t),t)$. Let s_k be a number, such that $|z(t)|\le s_k$ for almost all $t\in[0;t_k]$. Define the following functions:

$$\varphi_k(\sigma,t)=\begin{cases} \varphi(\sigma,t) \text{ if } |\sigma|\le s_k \\ \varphi(s_k,t) \text{ if } |\sigma|>s_k \\ \varphi(-s_k,t) \text{ if } |\sigma|<-s_k \end{cases}.$$

Consider a process $z_k(\cdot) = [\sigma_k(\cdot), \xi_k(\cdot)]$, for which $\xi_k(t) = \varphi_k(\sigma_k(t), t)$. This is a bounded process that satisfies all the quadratic constraints. Therefore, the sequence of processes $z_k(\cdot) = [\sigma_k(\cdot), \xi_k(\cdot)]$ is the required bounded continuation, which establishes that the system is minimally stable. By Theorem 2.6 the system is absolutely stable. The proof is complete.

If the function $\varphi(\sigma, t)$ is odd in σ, then, similarly to all the other results in this chapter, we can drop the requirement for the sequence θ_n to be nonnegative, replacing it with that of absolute convergence.

Theorem 4.10. *Assume that conditions 1), 2), and 3) of Theorem 4.9 are met and, in addition, the function $\varphi(\sigma, t)$ is odd in σ. Assume further that there exists an absolutely convergent series $\sum_{n=0}^{+\infty} \theta_n < 1/\zeta$, such that for all real values of ω*

$$\kappa^{-1} + \operatorname{Re}\left\{ W(i\omega)\left[1 - \sum_{n=0}^{+\infty} \theta_n e^{i\omega nT} \right] \right\} \ge \varepsilon > 0. \tag{4.4.12}$$

Then for all functions $\sigma(\cdot)$ and $\xi(\cdot)$, satisfying both (4.1.1) and (4.1.2), $\sigma(\cdot) \in L^2(0; +\infty)$ and, furthermore, there exists a positive constant λ, independent of the function $\alpha(\cdot)$, such that $\|\sigma(\cdot)\| \le \lambda \|\alpha(\cdot)\|$.

Proof. As in the proof of Theorem 4.9, define the quadratic forms $\mathcal{F}_1(\sigma_1, \xi_1)$ and $\mathcal{F}_1(\sigma_1, \xi_1, \sigma_2, \xi_2)$. The inequalities (4.4.10) and (4.4.11) still hold.

Define the additional quadratic form:

$$\mathcal{F}_3(\sigma_1, \xi_1, \sigma_2, \xi_2) = \zeta\sigma_1\xi_1 + \sigma_2\xi_1.$$

Lemma 2.10 implies the following inequality:

$$\int_0^t \mathcal{F}_3(\sigma(t), \xi(t), \sigma(t-T), \xi(t-T))dt \ge 0.$$

For the matrix $\Pi_3(\omega, n)$, we have the expression:

$$\Pi_3(\omega, n) = -\operatorname{Re}\left[\left(\zeta + e^{-i\omega nT} \right) W(i\omega) \right].$$

Therefore, the frequency condition takes the following form. There exist sequences θ_{n1}, θ_{n2}, and θ_{n3}, all with nonnegative terms, such that for all real values of ω

$$\text{Re}\left[\kappa^{-1}+W(i\omega)\right]\sum_{n=0}^{\infty}\theta_{1n}+\sum_{n=0}^{\infty}\theta_{2n}\,\text{Re}\left[\left(\zeta-e^{-i\omega nT}\right)W(i\omega)\right]$$

$$+\sum_{n=0}^{\infty}\theta_{3n}\,\text{Re}\left[\left(\zeta+e^{-i\omega nT}\right)W(i\omega)\right]\geq\varepsilon>0.$$

Again, we can set $\kappa^{-1}\sum_{n=0}^{\infty}\theta_{1n}+\zeta\left(\sum_{n=0}^{\infty}\theta_{2n}+\sum_{n=0}^{\infty}\theta_{3n}\right)=\kappa^{-1}$ without loss of generality. Define, as before, $\theta_n=\theta_{2n}-\theta_{3n}$ to conclude that this frequency condition is equivalent to (4.4.12). Minimal stability is verified by using the same function as in the proof of Theorem 4.9. The proof is concluded in the same way.

4.5 Linear Periodic Systems

Linear periodic systems received a considerable amount of attention in research literature going back to the fundamental work of Lyapunov [95]. The best known monograph on this subject is, undoubtedly, the two-volume classic by Yakubovich and Starzhinskii [168]. Some results can also be found in the books by Narendra and Taylor [109] and Venkatesh [138].

In this section we shall consider systems with the linear block given as before by the integral equation (4.1.1) and the function $\varphi(\sigma,t)$ given by

$$\varphi(\sigma,t)=P(t)\sigma(t)\,,\qquad\qquad(4.5.1)$$

where the matrix $P(t)$ is periodic with a period T. We shall assume that there exist constant matrices P and Q such that for all values of t

$$P*(t)Q-SP(t)\equiv0\,.\qquad\qquad(4.5.2)$$

Apart from the trivial possibility of $Q=S=0$, this condition appears to be rather restrictive. However, it is satisfied when $P(t)$ is a column vector and when it is a nonsingular square matrix satisfying some additional requirements. We shall consider these possibilities in two separate subsections. Finally, we shall consider a special case for which we shall prove a MIMO analogue of the Yakubovich criterion.

4.5.1 Case When P(t) Is a Column Vector

If the matrix $P(t)$ is a column vector, then the condition (4.5.2) holds for any pair of matrices Q and S such that $Q*=S$, where Q is a column vector of the same dimension as $P(t)$.

We assume that the column vector $P(t)$ satisfies an analogue of the sector condition, i.e., none of its components exceeds by absolute value a certain constant κ. Stated differently, $|P(t)|_\infty \leq \kappa$.

We have the following stability theorem.

Theorem 4.11. *Assume the following:*

1) The linear block (4.1.1) satisfies the regularity conditions;

2) The column vector $P(t)$ is periodic with a period T and $|P(t)|_\infty \leq \kappa$;

3) There exists a set of odd periodic functions $q_j(\omega)$, j=1, 2...p with a period $2\pi/T$ such that for all real values of ω

$$\kappa^1 I + \mathrm{Re}\left[EW(i\omega)\right] + \sum_{j=1}^{p} q_j(\omega) \mathrm{Im}\left[Q_j W(i\omega)\right] \geq \varepsilon I > 0, \qquad (4.5.3)$$

where E is a column vector of dimension p with all components equal to unity and Q_j is a column vector of dimension p with j^{th} component equal to unity and other components equal to zero.

Then for all functions $\sigma(\cdot)$ and $\xi(\cdot)$, satisfying both (4.1.1) and (4.1.2) with $\varphi(\sigma,t)$ given by (4.5.1), $\sigma(\cdot) \in L^2(0;+\infty)$ and, furthermore, there exists a positive constant λ, independent of the function $\alpha(\cdot)$, such that $\|\sigma(\cdot)\| \leq \lambda\|\alpha(\cdot)\|$.

Proof. Define a quadratic form:

$$\mathcal{F}_0\left(\sigma_1,\xi_1\right) = \xi_1^*\left(E\sigma_1 - \kappa^{-1}\xi_1\right).$$

It can be easily seen that

$$\mathcal{F}_0\left(\sigma_1(t),\xi_1(t)\right) > 0.$$

Furthermore, define the following set of p quadratic forms:

$$\mathcal{F}_j\left(\sigma_1,\xi_1,\sigma_2,\xi_2\right) = \xi_1^* Q_j \sigma_2 - \sigma_1^* Q_j^* \xi_2, j = 1, \ 2...p.$$

Condition (4.5.2) with $Q = Q_j$ and $S = Q_j^*$ implies the identity:

$$\mathcal{F}_j\left(\sigma(t),\xi(t),\sigma(t-nT),\xi(t-nT)\right) \equiv 0 \qquad (4.5.4)$$

Indeed

$$\xi*(t)Q_j\sigma(t-nT)-\sigma*(t)Q_j^*\xi(t-nT)$$
$$=\sigma*(t)P*(t)Q_j\sigma(t-nT)-\sigma*(t)Q_j^*P(t-nT)\sigma(t-nT)$$
$$=\sigma*(t)\left[P*(t)Q_j-Q_j^*P(t-nT)\right]\sigma(t-nT)$$
$$=\sigma*(t)\left[P*(t)Q_j-Q_j^*P(t)\right]\sigma(t-nT)$$
$$\equiv 0.$$

Note that (4.5.4) holds for the quadratic forms $-\mathcal{F}_j(\sigma_1,\xi_1,\sigma_2,\xi_2)$ as well.

Next, we compute the matrices $\Pi_j(\omega,n)$:

$$\Pi_0(\omega,n)=-\mathrm{Re}\left[\kappa^{-1}I+EW(i\omega)\right];$$
$$\Pi_j(\omega,n)=-\mathrm{Re}\left\{\left[W*(i\omega)Q_j^*-Q_jW(i\omega)\right]e^{-i\omega nT}\right\}$$
$$=2\,\mathrm{Im}\left[Q_jW(i\omega)\right]\sin\omega nT.$$

Therefore, the frequency condition takes the form: There exist sequences θ_{0n} with nonnegative terms and θ_{jn} such that for all real values of ω

$$\left\{\kappa^{-1}I+\mathrm{Re}\left[EW(i\omega)\right]\right\}\sum_{n=0}^{+\infty}\theta_{0n}$$
$$+\sum_{j=1}^{p}\left\{2\,\mathrm{Im}\left[Q_jW(i\omega)\right]\sum_{n=0}^{+\infty}\theta_{jn}\sin\omega nT\right\}\geq\varepsilon I>0 \tag{4.5.5}$$

Without loss of generality we can set $\sum_{n=0}^{\infty}\theta_{0n}=1$. Furthermore, set each θ_{jn} to be one half of the corresponding Fourier coefficient of the odd function $q_j(\omega)$. Hence the conditions (4.5.3) and (4.5.5) are equivalent.

Now let us replace (4.5.1) with the equation containing a parameter δ.

$$\varphi(\sigma,t)=\delta P(t)\sigma(t). \tag{4.5.6}$$

Since the frequency condition does not depend on δ, then by Lemma 2.7, for all values of δ the system is dichotomic, i.e., all bounded processes are stable. If $\delta=0$, all processes in the system are stable. If for any values of δ the system were to have any unstable processes, then for some critical value of δ it would have an unstable bounded process. This contradicts the fact of its dichotomy. Therefore, all processes in the system are stable for all values of δ, i.e., $\sigma(\cdot) \in L^2(0;+\infty)$. By Lemma 2.3, the estimate in the conclusion of the theorem is also valid. The proof is complete.

4.5.2 Case When P(t) Is a Nonsingular Square Matrix

Now we turn our attention to the case when $P(t)$ is a nonsingular $m \times m$ matrix. We shall assume that, in addition to (4.5.2), this matrix satisfies the MIMO analogue of the sector condition:

$$\operatorname{Re} P(t) - \kappa I \le 0. \tag{4.5.7}$$

Note that in this case (4.5.2) is satisfied if the matrix $P(t)$ is symmetric (set $Q=S=I$). It also holds if $P(t)=GH(t)$, where $H(t)$ is either symmetric (set $Q = S = G^{-1}$) or orthogonal (set $Q=G$ and $S = G^{-1}$).

A representation $P(t)=GH(t)$, where $H(t)$ is symmetric, can be found by the following method. Recall that every matrix can be decomposed into a product of a symmetric and an orthogonal matrix, i.e., we have $P(t)=U(t)H(t)$, where $H^2(t) = P*(t)P(t)$ and $U(t) = P(t)H^{-1}(t)$. If the matrix $U(t)$ turns out to be constant, set $G=U(t)$.

Let us illustrate this idea with a simple example. Let

$$P(t) = \begin{bmatrix} 1 & \sin t \\ \cos t & 1 \end{bmatrix}.$$

Performing the above computation, we find

$$U(t) = \begin{bmatrix} 0 & 1 \\ 1 & 0 \end{bmatrix}$$

and conclude that in this case there exists a desired representation.

Another way of characterizing the matrices satisfying (4.5.2) is given by the following statement.

Proposition 4.12. *Suppose that there exists a constant matrix Q such that*

$$\left[P'(t)P^{-1}(t) \right]*Q \equiv QP^{-1}(t)P'(t). \tag{4.5.8}$$

Then the matrix $S = P*(t)QP^{-1}(t)$ is constant.

Proof. Differentiation of the expression $S = P^*(t)QP^{-1}(t)$ yields

$$
\begin{aligned}
\frac{dS}{dt} &= \left[\frac{dP^*(t)}{dt}\right]QP^{-1}(t) + P^*(t)Q\frac{dP^{-1}(t)}{dt} \\
&= \left\{\left[\frac{dP^*(t)}{dt}\right]Q - P^*(t)QP^{-1}(t)P'(t)\right\}P^{-1}(t) \\
&= \left\{[P^*(t)]^{-1}\left[\frac{dP^*(t)}{dt}\right]Q - QP^{-1}(t)P'(t)\right\}P^{-1}(t) \\
&= \left\{\left[P'(t)P^{-1}(t)\right]^*Q - QP^{-1}(t)P'(t)\right\}P^{-1}(t) \equiv 0.
\end{aligned}
$$

Therefore, the matrix S is constant.

Hence, the determination of whether a given matrix $P(t)$ satisfies the condition (4.5.2) can be made by solving the equation (4.5.8) for the matrix Q. If a constant solution can be found, the question is answered affirmatively.

For the above example, this approach yields the following relationships for the components of the matrix Q: $q_{11} = q_{22}\tan t$ and $q_{12} = q_{21} - q_{22}\sec t$. Set $q_{11} = q_{22} = 0$ and $q_{12} = q_{21} = 1$ to obtain the desired constant matrix Q.

Let us state and prove the stability criterion for the systems of this type.

Theorem 4.13. *Assume the following:*

1) The linear block (4.1.1) satisfies the regularity conditions;

2) The matrix $P(t)$ is nonsingular, periodic with a period T, and satisfies the condition (4.5.7);

3) There exist constant matrices Q and S, such that (4.5.2) holds for all values of t;

4) There exists an odd periodic function $q(\omega)$ with a period $2\pi/T$ such that for all real values of ω

$$
\operatorname{Re}\left\{\kappa^{-1}I + W(i\omega) + iq(\omega)[QW(i\omega) - W^*(i\omega)S]\right\} \geq \varepsilon I > 0. \qquad (4.5.9)
$$

Then for all functions $\sigma(\cdot)$ and $\xi(\cdot)$, satisfying both (4.1.1) and (4.1.2) with $\varphi(\sigma,t)$ given by (4.5.1), $\sigma(\cdot) \in L^2(0;+\infty)$ and, furthermore, there exists a positive constant λ, independent of the function $\alpha(\cdot)$, such that $\|\sigma(\cdot)\| \leq \lambda\|\alpha(\cdot)\|$.

Proof. Define a quadratic form:

$$
\mathcal{F}_1(\sigma_1,\xi_1) = \xi_1^*\left(\sigma_1 - \kappa^{-1}\xi_1\right).
$$

Condition (4.5.7) implies that

$$\mathcal{F}_1\left(\sigma(t),\xi(t)\right) \ge 0.$$

Indeed

$$\mathcal{F}_1\left(\sigma(t),\xi(t)\right) = \xi^*(t)\left(P^{-1}(t) - \kappa^{-1}I\right)\xi(t). \tag{4.5.10}$$

Condition (4.5.7) means that none of the eigenvalues of the matrix $P(t)$ exceed κ by absolute value. This implies that absolute values of all the eigenvalues of the matrix $P^{-1}(t)$ are at least κ^{-1}, which means that the expression in the right-hand side of (4.5.10) is nonnegative.

Define two more quadratic forms:

$$\mathcal{F}_2\left(\sigma_1,\xi_1,\sigma_2,\xi_2\right) = \xi_1^*Q\sigma_2 - \sigma_1^*S\xi_2;$$
$$\mathcal{F}_3\left(\sigma_1,\xi_1,\sigma_2,\xi_2\right) = \sigma_2^*S\xi_1 - \xi_2^*Q\sigma_1.$$

Let us verify that condition (4.5.2) implies the identity

$$\mathcal{F}_j\left(\sigma(t),\xi(t),\sigma(t-nT),\xi(t-nT)\right) \equiv 0 \,,\ j{=}2,3. \tag{4.5.11}$$

Indeed, for $j = 2$ we have

$$\xi^*(t)Q\sigma(t-nT) - \sigma^*(t)S\xi(t-nT)$$
$$= \sigma^*(t)P^*(t)Q\sigma(t-nT) - \sigma^*(t)SP(t-nT)\sigma(t-nT)$$
$$= \sigma^*(t)\left[P^*(t)Q - SP(t-nT)\right]\sigma(t-nT)$$
$$= \sigma^*(t)\left[P^*(t)Q - SP(t)\right]\sigma(t-nT)$$
$$\equiv 0.$$

The proof for $j = 3$ is similar.

Note that the identity (4.5.11) holds for the quadratic forms $-\mathcal{F}_j(\sigma_1,\xi_1,\sigma_2,\xi_2)$ as well.

Next, we compute the matrices $\Pi_j(\omega,n)$:

$$\Pi_1(\omega,n) = -\text{Re}\left[\kappa^{-1}I + W(i\omega)\right];$$
$$\Pi_2(\omega,n) = -\text{Re}\left\{[W^*(i\omega)S - QW(i\omega)]e^{-i\omega nT}\right\};$$
$$\Pi_3(\omega,n) = -\text{Re}\left\{[QW(i\omega) - W^*(i\omega)S]e^{i\omega nT}\right\}.$$

Therefore, after some algebraic manipulations the frequency condition takes the following form. There exist sequences θ_{1n} with nonnegative terms, θ_{2n}, and θ_{3n}, such that for all real values of ω

$$\mathrm{Re}\left\{\left[\kappa^{-1}I+W(i\omega)\right]\sum_{n=0}^{+\infty}\theta_{1n}+Y(i\omega)\sum_{n=0}^{+\infty}\left(\theta_{2n}e^{-in\omega T}-\theta_{3n}e^{in\omega T}\right)\right\}\geq\varepsilon I>0,\quad(4.5.12)$$

where $Y(i\omega)=W*(i\omega)S-QW(i\omega)$.

If (4.5.9) holds, then (4.5.12) holds as well. Indeed, we can expand the function $q(\omega)$ in a Fourier series. Since this function is assumed to be odd, the expansion has the form $q(\omega)=\sum_{n=0}^{+\infty}\theta_n\sin n\omega T$. Set $\sum_{n=0}^{\infty}\theta_{1n}\equiv1$, $\theta_{2n}=\theta_{3n}=\theta_n$.

We can now complete the proof in the same way as in Theorem 4.7.

Note that if we set $Q=S=0$, the frequency condition takes the form:

$$\mathrm{Re}\left[\kappa^{-1}I+W(i\omega)\right]\geq\varepsilon I>0.$$

Clearly, this is a MIMO analogue of the circle criterion, which does not make use of any special properties of the system, such as linearity or periodicity.

4.5.3 Yakubovich Criterion and Its MIMO Analogue

Suppose that the matrix $P(t)$ is symmetric, which also includes the SISO case. Then we can set $Q=S=I$ and the frequency condition can be rewritten as follows:

$$\mathrm{Re}\left\{\left[\kappa^{-1}I+W(i\omega)\right]\left[1+iq(\omega)\right]\right\}\geq\varepsilon I>0.\quad(4.5.13)$$

This is a MIMO analogue of Yakubovich's result [166]. It is important to note that it has the multiplier form. Therefore, in the SISO case, the system can be investigated by means of the Lipatov plot.

Consider once again the system with the linear block described by the transfer function

$$W(s)=\frac{s^2}{\left[(s+0.5)^2+0.81\right]\left[(s+0.5)^2+1.21\right]}.$$

Let $\kappa=11$. Set $q(\omega)=-0.7\sin\omega T$. Figure 4.6 shows the Lipatov plot for $T=0.55\pi$, which allows us to conclude that the system is stable.

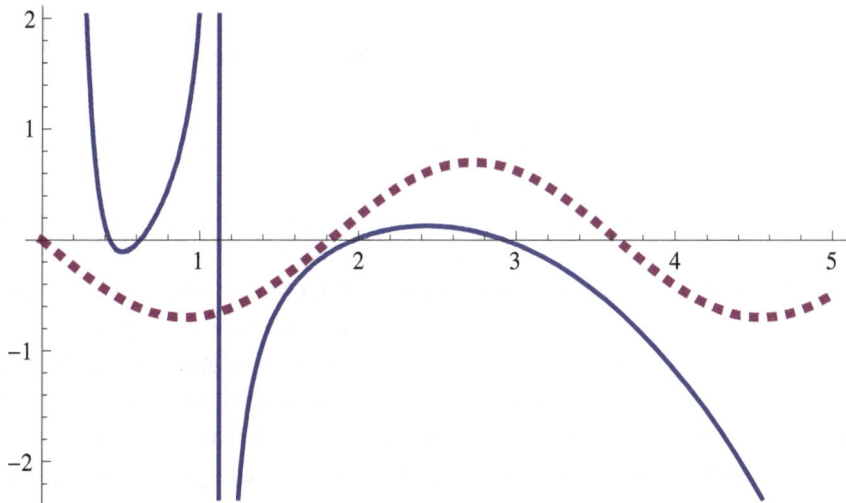

Fig. 4.6 Lipatov plot for $T=0.55\pi$. Function $\Phi(\omega)$ is shown as a solid line and function $\Psi(\omega)$ is shown as a broken line

Now let us use the same function $q(\omega)$ with $T=0.63\pi$. The Lipatov plot is shown in Fig. 4.7, and we can once again conclude that the system is stable.

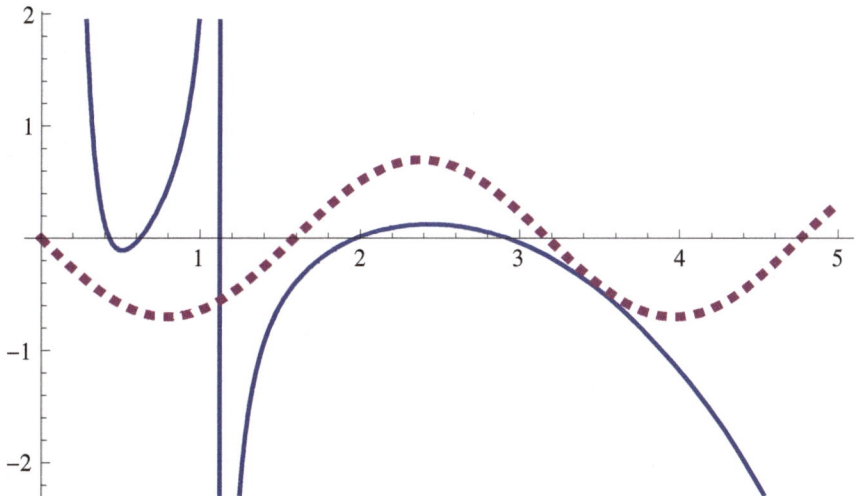

Fig. 4.7 Lipatov plot for $T=0.63\pi$. Function $\Phi(\omega)$ is shown as a solid line and function $\Psi(\omega)$ is shown as a broken line

As another example, let us consider a generalized damped Hill equation, i.e., let the linear block be defined by the transfer function:

$$W(s) = \frac{1}{s^2 + as + b}.$$

In this case the plot of the function $\Phi(\omega)$ consists of only one branch with the asymptote at $\omega=0$. This branch intersects the ωaxis in points:

$$\omega_{1,2} = \frac{\sqrt{-a^2 + 2b + \zeta \mp \sqrt{a^4 + \zeta^2 - 2a^2(2b+\zeta)}}}{\sqrt{2}}. \qquad (4.5.14)$$

We can now observe that we can always find the desired function $q(\omega)$ with the period greater than $2\omega_2$. Hence the system is stable if $T < \pi / \omega_2$.

If either of the two radicands in the numerator in (4.5.14) is negative, the curve $\Phi(\omega)$ does not intersect the horizontal axis and the circle criterion applies.

Let us apply this result to the standard benchmark – the slightly damped Mathieu equation with $\zeta = 0.1$. In this case, $p(t) = \alpha - 2q\cos 2t$, which yields $\delta = \alpha - 2q$ and $\kappa = 4q$. For $T = \pi$, the above criterion, combined with the circle criterion yields the stability region in the plane of the parameters α and q shown in Fig. 4.8.

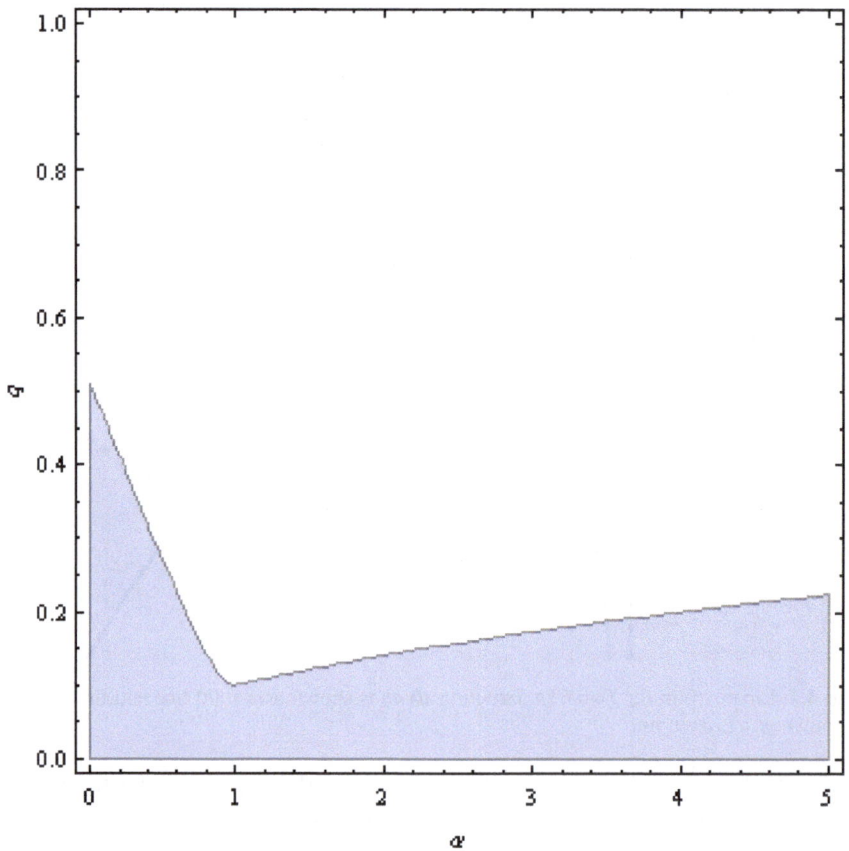

Fig. 4.8 Stability zone for the slightly damped Mathieu equation

4.6 SISO Systems with Parametric Class of Nonlinearities

In this section we shall extend the results obtained for the SISO linear periodic systems to a slightly more general case. As in the previous two sections, it will be assumed that the nonlinearity $\varphi(\sigma,t)$ satisfies the sector condition

$$0 \leq \frac{\varphi(\sigma,t)}{\sigma} \leq \kappa. \tag{4.6.1}$$

Furthermore, we assume that there exists a real quadratic form $\mathcal{G}(u,v)$ such that for all values of t

$$\left| \sigma_1(t)\varphi(\sigma_2(t),t) - \sigma_2(t)\varphi(\sigma_1(t),t) \right| \leq \mathcal{G}\big(\sigma_1(t),\varphi(\sigma_1(t),t)\big)$$
$$+ \mathcal{G}\big(\sigma_2(t),\varphi(\sigma_2(t),t)\big). \tag{4.6.2}$$

Of particular interest is the case with $\mathcal{G}(u,v) = \gamma v\left(u - \kappa^{-1}v\right)$. If $\gamma \neq 0$, then (4.6.2) holds if and only if $\varphi(\sigma,t)$ is a linear function of σ. If $\gamma = +\infty$, then this inequality holds for all functions.

The stability criterion is stated as follows.

Theorem 4.14. *Assume the following:*

1) The linear block (4.1.1) *satisfies the regularity conditions;*

2) The function $\varphi(\sigma,t)$ *is continuous in each argument and periodic in t with a period T,*

3) There exists a real quadratic form $\mathcal{G}(u,v)$, *such that* (4.6.2) *holds for all values of t;*

4) There exists an odd periodic function $q(\omega)$ *with a period* $2\pi/T$ *such that for all real values of* ω

$$\mathrm{Re}\left[\kappa^{-1} + W(i\omega) - \mathcal{G}(-W(i\omega),1) - iq(\omega)W(i\omega) \right] \geq \varepsilon > 0. \tag{4.6.3}$$

Then for all functions $\sigma(\cdot)$ *and* $\xi(\cdot)$, *satisfying both* (4.1.1) *and* (4.1.2) *with* $\varphi(\sigma,t)$ *given by* (4.5.1), $\sigma(\cdot) \in L^2(0;+\infty)$ *and, furthermore, there exists a positive constant* λ, *independent of the function* $\alpha(\cdot)$, *such that* $\|\sigma(\cdot)\| \leq \lambda \|\alpha(\cdot)\|$.

Proof. Define a quadratic form:

$$\mathcal{F}_1(\sigma_1,\xi_1) = \xi_1\left(\sigma_1 - \kappa^{-1}\xi_1\right).$$

Condition (4.6.1) implies that

$$\mathcal{F}_1\left(\sigma(t),\xi(t)\right)\ge 0.$$

Next, define two more quadratic forms:

$$\mathcal{F}_2(\sigma_1,\xi_1,\sigma_2,\xi_2)=\mathcal{G}(\sigma_1,\xi_1)+\mathcal{G}(\sigma_2,\xi_2)+\sigma_1\xi_2-\sigma_2\xi_1,$$
$$\mathcal{F}_3(\sigma_1,\xi_1,\sigma_2,\xi_2)=\mathcal{G}(\sigma_1,\xi_1)+\mathcal{G}(\sigma_2,\xi_2)+\sigma_2\xi_1-\sigma_1\xi_2.$$

The inequality (4.6.2) implies

$$\mathcal{F}_j\left(\sigma(t),\xi(t),\sigma(t-nT),\xi(t-nT)\right)\ge 0,\quad j=2,3.\qquad(4.6.4)$$

Compute the matrices $\Pi_j(\omega,n)$ to obtain

$$\Pi_1(\omega,n)=-\mathrm{Re}\left[\kappa^{-1}+W(i\omega)\right];$$
$$\Pi_2(\omega,n)=-2\,\mathrm{Re}\left[-\mathcal{G}(-W(i\omega),1)+iW(i\omega)\sin\omega nT\right];$$
$$\Pi_3(\omega,n)=-2\,\mathrm{Re}\left[-\mathcal{G}(-W(i\omega),1)-iW(i\omega)\sin\omega nT\right].$$

The frequency condition takes the following form. There exist se-
quences θ_{1n}, θ_{2n}, and θ_{3n}, all with nonnegative terms such that for all real values
of ω

$$\mathrm{Re}\left\{\left[\kappa^{-1}+W(i\omega)\right]\sum_{n=0}^{+\infty}\theta_{1n}-Z_-(\omega,n)\sum_{n=0}^{+\infty}\theta_{2n}-Z_+(\omega,n)\sum_{n=0}^{+\infty}\theta_{3n}\right\}\ge\varepsilon>0,\qquad(4.6.5)$$

where

$$Z_-(\omega,n)=\mathcal{G}(-W(i\omega),1)-iW(i\omega)\sin\omega nT;$$
$$Z_+(\omega,n)=\mathcal{G}(-W(i\omega),1)+iW(i\omega)\sin\omega nT.$$

If the condition (4.6.3) holds, so does (4.6.5). To see this, expand the function
$q(\omega)$ in a Fourier series. Since this function is assumed to be odd, the expansion
has the form $q(\omega)=\sum_{n=0}^{+\infty}\theta_n\sin n\omega T$. Choose the sequences θ_{1n}, θ_{2n}, and θ_{3n}
such that

$$\frac{\sum_{n=0}^{+\infty}(\theta_{2n}+\theta_{3n})}{\sum_{n=0}^{+\infty}\theta_{1n}}=1$$

and

$$\theta_n = \frac{\sum_{n=0}^{+\infty}(\theta_{2n}-\theta_{3n})}{\sum_{n=0}^{+\infty}\theta_{1n}}.$$

Minimal stability can be verified by using the same functions as in the proof of Theorem 4.5. The proof is concluded in the same way.

Let us turn our attention to the special case of $\mathcal{G}(u,v) = \gamma v\left(u - \kappa^{-1}v\right)$. After some algebraic manipulations, the condition (4.6.3) takes the form:

$$\mathrm{Re}\left\{\left[\kappa^{-1}+W(i\omega)\right]\left[1+\gamma+iq(\omega)\right]\right\} \geq \varepsilon > 0. \tag{4.6.6}$$

Note that this condition has the multiplier form and, hence, is amenable to analysis by means of a Lipatov plot. If $\gamma = 0$, it reduces to (4.5.13), which is to be expected since only the linear time-periodic functions satisfy (4.6.2) in this case. It is easy to see that if a function $q(\omega)$ can be found for $\gamma = 0$, one can also be found for any positive value of γ. As γ approaches infinity, the condition (4.6.6) tends pointwise to the circle criterion as the class of functions satisfying (4.6.2) expands to include all functions satisfying the sector condition (4.6.1).

4.7 Concluding Remarks

Unlike the case of the stationary nonlinearities, stability theory for time-periodic systems is not yet very well developed. Results presented in this chapter suggest that it may proceed along the same lines. In fact, the only difference between the two types of frequency conditions is the use of the Fourier series instead of integrals.

The last fact means that stability multipliers are periodic in the frequency domain. Therefore, their matrix realization may not exist, which precludes their numerical implementation via matrix inequalities. It is possible, however, to use the geometric method of Lipatov, but it requires a certain amount of guesswork in selecting the trigonometric polynomial. Therefore, one of the open problems is to find an algorithm that will eliminate or reduce this guesswork.

It also must be noted that stability criteria for certain types of systems do not have the multiplier form, which precludes the use of Lipatov plots. A possible approach in these cases may be an iterative procedure developed by Skorodinskii [132, 133], which does not rely on the multiplier form of the frequency conditions. A cursory review of his works suggests that these methods may be applicable if the Fourier series are truncated and the problem is reduced to finding the optimal set of the coefficients.

I hope that results presented in this chapter will give the reader a good starting point for further research of stability problems for time-periodic systems.

A
Appendix: Prerequisites

A.1 Measure

The concept of measure generalizes the familiar notions of length, area, and volume. A reader, familiar with stochastic processes, may have also encountered this term, since probability is defined as a certain measure.[1]

Intuitively speaking, we define the measure of a set $A \subset \mathbb{R}$ to be a real function $\mu(A)$, such that the following conditions (we call them the axioms of measure) are satisfied:

1) $\mu(\varnothing) = 0$.
2) If the sets A_j, $j \in \mathbb{N}$, are pairwise disjoint, then

$$\mu\left(\bigcup_{j=1}^{\infty} A_j\right) = \sum_{j=1}^{\infty} \mu(A_j).$$

Unfortunately, it is impossible to define this function in such a way that would enable us to compute it for every subset of $\mathbb{R}$. For this reason we confine our attention only to sets for which such computation is possible. We term such sets measurable.

If for every measurable set A, it is true that $\mu(A) \geq 0$, the measure is called nonnegative, otherwise we call it signed. If for every measurable set A (including $\mathbb{R}$), $\mu(A) < \infty$, the measure is called finite. An example of a finite measure is probability: Let $\mu(A)$ be the probability that the output of some random number generator belongs to A. Then $\mu(\mathbb{R}) = 1$.

An important role in analysis is played by the Lebesgue measure, $\mu_L(A)$, which can be axiomatically defined by requiring that $\mu_L(a;b) = b - a$, i.e., the Lebesgue measure of an interval or a segment is its length. It can be shown (See, for example,

---

[1] Readers who studied probability without ever hearing about measure may now find themselves in the position of Monsieur Jourdain from Moliere's *Le Bourgeois Gentilhomme*, who suddenly found out that he was speaking prose all his life without knowing it!

[129]) that this condition, together with the axioms of measure, uniquely specifies the function $\mu_L(A)$ for any set A for which the Lebesgue measure can be defined. Clearly, the Lebesgue measure is nonnegative, but not finite.

Obviously, the Lebesgue measure of a set consisting of only one point is zero. This implies that the Lebesgue measure of a countable set is also zero because any such set consists only of isolated points. The converse is false: There exists an uncountable set of Lebesgue measure zero.[2] Also, as noted before, there are sets for which Lebesgue (or any other) measure cannot be defined.[3]

If a certain property holds for all real numbers except those that belong to set of Lebesgue measure zero, it is customary to say that this property holds almost everywhere (commonly abbreviated a.e.), and we will do so from now on.

Another concept of measure that plays an important role is that of Stieltjes, which is defined as follows. Let $f(t)$ be a nondecreasing lower-semicontinuous function[4] of a scalar argument t. Define

$$\mu(a;b) \triangleq f(b) - f(a+0);$$
$$\mu[a;b] \triangleq f(b+0) - f(a);$$
$$\mu(a;b] \triangleq f(b+0) - f(a+0);$$
$$\mu[a;b) \triangleq f(b) - f(a).$$

Similarly to the Lebesgue measure, it can be shown that these conditions, together with the axioms of measure, uniquely specify the function $\mu(A)$. The measure thus defined is called the Lebesgue-Stieltjes measure. It is easy to see that the Lebesgue-Stieltjes measure is nonnegative. The function $f(t)$ is called the generating function for the measure, which we denote μ_f. It can be shown that every nonnegative measure can be generated by a nondecreasing lower-semicontinuous function. Note that if we set $f(t) = t$, we obtain the Lebesgue measure.

If we relax the assumption that the function $f(t)$ is nondecreasing by requiring instead for it to be a difference of two nondecreasing functions (functions that satisfy this requirement are said to be of bounded variation[5]), the above formulas define a signed measure. Furthermore, every signed measure can be generated by a function of bounded variation.

---

[2] Interested readers are encouraged to look up information on the so-called Cantor tertiary set.

[3] Interested readers may consult the book by Schilling [129] for more details.

[4] A function $f(t)$ is called lower-semicontinuous if it is continuous from below, that is for any sequence of numbers $t_n < t_0$ and converging to the point t_0, the sequence $f(t_n)$ converges to $f(t_0)$. Notation $f(t+0)$ will be used to denote a jump that the function $f(t)$ may undergo at the point t_0.

[5] This is not the rigorous definition, but it will suffice for our purposes. An example of a function that is not of bounded variation is $t\sin(1/t)$.

Unlike the Lebesgue measure, the Lebesgue-Stieltjes measure of a set consisting of only one point is not necessarily zero. Suppose that $f(t) = 0$ for $t \leq 0$ and $f(t) = 1$ for $t > 0$. Then from the second axiom of measure we have $\mu_f[0;1) = \mu_f(\{0\}) + \mu_f(0;1)$, which implies that $\mu_f(\{0\}) = 1$, since $\mu_f[0;1) = 1$ and $\mu_f(0;1) = 0$.

A.2 Lebesgue Integral

The reader undoubtedly remembers how the concept of integration was introduced in a basic calculus course. The interval of integration was partitioned and the so-called Riemann sums were defined. Then, the Fundamental Theorem of Calculus was proved, which asserted that the derivative with respect to the upper limit of the integral of the continuous function is the integrand.

This concept of integration is sufficient for all practical purposes. However, it has a few theoretical flaws. One of them is that the class of functions that can be integrated this way is "too narrow." What exactly we mean by this statement will become clear as we proceed.

In order to address these theoretical issues, Lebesgue introduced a more flexible concept of integration that involves partitioning the range, rather than the domain of a function.

First, we need to introduce the concept of a measurable function. We say that the function $f(t)$ is measurable if the set $A = \{t : f(t) < c\}$ is measurable for every real number c. Again, this is not the formal definition, but it will suffice for our purposes. Of course, whether a given function is measurable depends on the measure we use. However, the one used most frequently is the Lebesgue measure, and we will use it as well without stating the name explicitly. Practically speaking, any function one can reasonably think of is measurable.

Next, we consider a class of the so-called simple functions. A function $f(t)$ is called simple if its range is either a finite or a countable set. Let the values of this function be denoted by y_j. The simple function $f(t)$ is measurable if and only if all sets $A_j = \{t : f(t) = y_j\}$ are measurable. An arbitrary function $f(t)$ is measurable if and only if it is a limit of a uniformly convergent[6] sequence of simple measurable functions.

The integral of a simple function $f(t)$ with respect to a measure μ is defined by

$$\int_A f(t) d\mu(t) = \sum_j y_j \mu(A_j), \quad A_j = \{t \in A : f(t) = y_j\}.$$

---

[6] Roughly speaking, uniform convergence of a sequence of functions $f_n(t)$ to the limit function $f(t)$ means that the rate of convergence does not depend on t.

The expression in the right-hand side of the integral may be either a finite sum or an infinite series. If the latter converges, the function $f(t)$ is called Lebesgue integrable or simply integrable. If the sum is finite, then the measurability of the function $f(t)$ is necessary and sufficient for it to be integrable.

This definition roughly means that for each of the possible values of the function $f(t)$, we compute the measure of the set on which the function takes this value and then add the results of such computations.

An arbitrary function $f(t)$ is integrable if it is a limit of a uniformly convergent sequence $f_n(t)$ of simple integrable functions. If so, we write

$$\int_A f(t)d\mu(t) = \lim_{n \to \infty} \int_A f_n(t)d\mu(t).$$

It can be shown that this limit always exists for any uniformly convergent sequence $f_n(t)$ of simple integrable functions and, furthermore, for the given function $f(t)$, it does not depend on a choice of the sequence $f_n(t)$. Furthermore, if the function $f(t)$ is bounded on the set A, its measurability is necessary and sufficient for it to be integrable.

If the Lebesgue measure is used, which is the most common case, the integral we have just defined is called the Lebesgue integral, and dt is often written instead of $d\mu(t)$. From now on, all the integrals will be understood in this sense unless it is explicitly stated that some other measure is being used. If the set A is a segment, we may resort to the familiar notation involving lower and upper limits of integration.

It is important to note that if the function $f(t)$ is integrable using the familiar Riemann sum procedure (we call such functions Riemann integrable), then both approaches give the same values. All the integration techniques learned in a standard calculus course are applicable to Lebesgue integrals as well.

There is a theorem that states that a function is Riemann integrable on a certain interval if and only if it is continuous a.e. on this interval. A sequence of uniformly convergent Riemann integrable functions is also Riemann integrable. Without the requirement for the convergence to be uniform, this statement is false. From the theoretical point of view, this restriction turns out to be too severe, and this is the reason for using the Lebesgue approach. From the practical point of view, however, there is relatively little loss in thinking about integrals in terms of the ordinary Riemann definition.

A.3 Lebesgue-Stieltjes Integral

Another definition of integral that is sometimes used in this book is that of Lebesgue-Stieltjes. As the term suggests, this definition involves the use of Lebesgue-Stieltjes measures and functions that generate them.

Let $\vartheta(t)$ be a function of bounded variation that generates the signed measure μ_ϑ. Then, we introduce the following integral expression:

$$\int_A f(t)d\vartheta(t) = \int_A f(t)d\mu_\vartheta(t).$$

The integral expression in the left-hand side is called the Stieltjes integral. Because we defined it using the concept of measure, we call it the Lebesgue-Stieltjes integral. It is thus distinguished from the original Stieltjes approach which used the procedure similar to the Riemann sums. The latter proved to be unsatisfactory, because it requires continuity of the function $f(t)$ on the interval of integration.

The reason for using the Stieltjes notation is that sometimes we need to emphasize certain properties of the generating function as opposed to the measure. Let us elaborate this idea a little further. In order to do this, we need to define some classes of functions.

From the standard calculus course, it is known that if the function $f(t)$ is continuous on a segment $[a;b]$, then, the following equality holds for every $t \in [a;b]$:

$$\int_a^t f'(t)dt = f(t) - f(a),$$

where the integral is understood in the sense of Riemann.

When the Lebesgue integral is considered, this statement is no longer true in general. In order to make it true, a narrower class of functions, called absolutely continuous, is introduced. A function $f(t)$ is called absolutely continuous on a segment $[a;b]$ if for every $\varepsilon > 0$ there exists $\delta > 0$ such that

$$\sum_{j=1}^n \left| f(\alpha_j) - f(\beta_j) \right| < \varepsilon$$

for every finite collection of intervals $(\alpha_j; \beta_j) \subset [a;b]$ with

$$\sum_{j=1}^n \left| \alpha_j - \beta_j \right| < \delta.$$

The crucial property of absolutely continuous functions is that they are differentiable a.e. This statement is generally false for continuous functions. In fact, there are functions continuous on a segment but not differentiable at any point of this segment![7] Another important property of absolutely continuous functions is that they are of bounded variation.

---

[7] Such functions are often called nowhere-differentiable. See, for example, [99, p. 113].

Furthermore, it is possible to show that for Lebesgue integrals, the equality

$$\int\limits_a^t f'(t)dt = f(t) - f(a)$$

completely characterizes the class of absolutely continuous functions. In addition, the Lebesgue integral of any integrable function over a segment is an absolutely continuous function of the upper limit of integration.

Let us now introduce the concept of a saltus function constructed as follows. Define within a segment $[a;b]$ a finite or a countable set of points denoted by t_j. Suppose further that each point t_j has a finite number h_j associated with it, and furthermore, $\sum_j h_j < \infty$. Let

$$h(t) = \sum_{t_j < t} h_j.$$

The function $h(t)$ thus defined is called a saltus function.

For every function $f(t)$ of bounded variation, we have $f(t) = g(t) + h(t)$, where $g(t)$ is a continuous function and $h(t)$ is a saltus function. If the function $g(t)$ is not absolutely continuous, we define a function $\phi(t)$ by

$$\phi(t) = \int\limits_a^t g'(t)dt .$$

Since the function $g(t)$ is of bounded variation, it is differentiable a.e., which means that the integral in the right-hand side exists.

It can be shown that the derivative of the difference $g(t) - \phi(t)$ is zero a.e. A function with the derivative equal to zero a.e. is called singular. It is possible that the derivative of a singular function at points where it is not zero may not exist.

Therefore, every function of bounded variation can be represented as a sum of three functions (we call them components): an absolutely continuous function, a saltus function, and a singular function. This is called the canonical decomposition.

With this in mind, let us consider some special cases of the Lebesgue-Stieltjes integral. If the function $\vartheta(t)$ is absolutely continuous, it can be shown that

$$\int\limits_a^b f(t)d\vartheta(t) = \int\limits_a^b f(t)\vartheta'(t)dt .$$

In other words, the Lebesgue-Stieltjes integral reduces to the Lebesgue integral.

Next, consider the case when $\vartheta(t)$ is a saltus function having jumps equal to h_j at points t_j. Then we find

$$\int_a^b f(t)\,d\vartheta(t) = \sum_j f(t_j)h_j,$$

i.e., the integral degenerates into a finite sum or infinite series.

Putting the above considerations together, we conclude that if the function $\vartheta(t)$ has only the absolutely continuous and the saltus components, the Lebesgue-Stieltjes integral can be reduced to the sum of the Lebesgue integral and the series. If the singular component is present, such reduction is impossible. This fact will play an important role when we compare some of the stability criteria proved in this book with those obtained in earlier papers on the subject.

A.4 Norms and L^p Spaces

The reader has most likely encountered the notion of a vector space either in a linear algebra or state-space methods course. It might have been mentioned in passing that certain functions form various vector spaces. This concept will now be discussed in some more detail.

One of the concepts often associated with a vector space is the norm, which is a natural generalization of the length of a vector. The function $\rho(x)$ of a vector x is called a norm if it satisfies the following three properties:

1) For any vector x, $\rho(x) \geq 0$ with $\rho(x) = 0$ if and only if $x=0$;

2) For any real or complex number α and any vector x, $\rho(\alpha x) = |\alpha|\rho(x)$;

3) For any two vectors x and y, $\rho(x+y) \leq \rho(x) + \rho(y)$.

For usual finite-dimensional vector spaces $\mathbb{R}^m$, the most commonly used norm is Euclidean, denoted by the single bars and defined by

$$|x| = \sqrt{\sum_{j=1}^m x_j^2}\ .$$

An important point about the Euclidean norm is that it can also be defined using the product of a vector with itself:

$$|x| = \sqrt{x * x}\ ,$$

where the star denotes the vector transpose operation.

Another vector norm that is sometimes used is the maximum of the absolute values of all the components. It is denoted by single bars with the index infinity:

$$|x|_\infty = \max_j |x_j| .$$

The space $L^1(a;b)$ is defined as a class of vector functions in $\mathbb{R}^m$ whose components have finite Lebesgue integral on the interval $(a;b)$ and either endpoint can be either positive or negative infinity. The norm in this vector space is given by

$$\|x(\cdot)\|_1 = \int_a^b |x(t)| dt .$$

The subscript next to double bars is used to indicate that we are considering the $L^1(a;b)$ function space. We are using the dot in place of the argument of the function to indicate that we are considering this function itself as an element of a vector space as opposed to the value of a specific function at a specific point, which is the case in the right-hand side.

In a similar way we define the spaces $L^p(a;b)$ for $p > 1$ with the norm given by

$$\|x(\cdot)\|_p = \sqrt[p]{\int_a^b |x(t)|^p \, dt} .$$

In addition, we define the space $L^\infty(a;b)$ to include all functions with absolute values not exceeding a certain finite number except on a set of measure zero. This number is called an essential supremum and will be defined to be the L^∞-norm of the function:

$$\|x(\cdot)\|_\infty = \operatorname{ess\,sup} |x(t)| .$$

Similarly, the norm in each of the L^p spaces is called the L^p-norm. The L^2-norm is usually called the Euclidean norm of a (possibly vector) function. Since it will occur most frequently, we will often omit the index and use the notation $\|\cdot\|$. This norm must be distinguished from the Euclidean norm of a value $x(t)$ of a vector function in a finite dimensional space, which, as stated above, is denoted by single bars.

If the interval $(a;b)$ is finite, then $p < q$ implies that $L^p(a;b) \subset L^q(a;b)$. The statement is not true if either endpoint is infinite. However, the following statement is true for any interval, finite or infinite: $L^1 \cap L^\infty \subset L^p$ for any $p > 1$.

Let us consider some elementary examples.

1) The function $f_1(t) \equiv 1$ belongs to the space $L^\infty(0;\infty)$ but not to any other of the $L^p(0;\infty)$ spaces.

2) The function

$$f_2(t) = \frac{1}{1+t}$$

belongs to both $L^2(0;\infty)$ and $L^\infty(0;\infty)$ but not to the $L^1(0;\infty)$ space.

3) The function

$$f_3(t) = \frac{1}{1+t} \frac{1+\sqrt[4]{t}}{\sqrt[4]{t}}$$

belongs to the $L^2(0;\infty)$ space but not to either the $L^1(0;\infty)$ or $L^\infty(0;\infty)$ space.

4) The function $f_4(t) = a^{-t}$ belongs to both the $L^1(0;\infty)$ and $L^\infty(0;\infty)$ spaces. Hence, it also belongs to all the other $L^p(0;\infty)$ spaces with $p \geq 1$, which can be easily verified.

5) The function

$$f_5(t) = \frac{1}{1+t^2} \frac{1+\sqrt[4]{t}}{\sqrt[4]{t}}$$

belongs to both the $L^1(0;\infty)$ and $L^2(0;\infty)$ space but not to $L^\infty(0;\infty)$.

6) The function

$$f_6(t) = \frac{1}{1+t^2} \frac{1+\sqrt{t}}{\sqrt{t}}$$

belongs to the $L^1(0;\infty)$ space, but not to either $L^2(0;\infty)$ or $L^\infty(0;\infty)$.

More information about L^p spaces and corresponding norms can be found in the book by Desoer and Vidyasagar [44].

Of all the L^p spaces, the most crucial role throughout this book is played by the L^2. The reason for this lies in the fact that $L^2(a,b)$ is the only L^p space in which we can define the product of two vectors:

$$\langle x(\cdot), y(\cdot) \rangle = \int_a^b |x*(t)y(t)| dt .$$

The norm (which for this reason will be called the Euclidean norm) can then be defined by

$$\|x(\cdot)\|_2 = \sqrt{\langle x(\cdot), x(\cdot) \rangle} \,.$$

Furthermore, the most crucial properties of the Fourier transform, which is discussed in the next section, namely the convolution property and the Plancherel theorem, are valid only for the functions that belong to the $L^2(-\infty, +\infty)$ space.

A.5 Two Facts about Fourier Transforms

Assuming that the function $f(t)$ is defined only for $t \geq 0$ (usually the case), its Fourier transform $\hat{f}(i\omega)$ can be obtained from its Laplace transform by substituting $i\omega$ for the Laplace variable s. More formal definition, which does not require the restriction of $t \geq 0$, is

$$\hat{f}(i\omega) = \int_{-\infty}^{+\infty} f(t)e^{i\omega t} dt \,.$$

If the two functions $f(\cdot)$ and $g(\cdot)$ both belong to the space $L^2(0, \infty)$, then the following two properties are valid and play a crucial role throughout the book.

The first property is the convolution theorem. Define the convolution of these two functions by

$$h(t) = f(t) * g(t) = \int_0^t f(\tau)g(t-\tau)d\tau \,.$$

The Fourier transform of the convolution is the product of the Fourier transforms of these two functions:

$$\hat{h}(i\omega) = \hat{f}(i\omega)\hat{g}(i\omega) \,.$$

The second property is known as the Plancherel[8] theorem. It is stated as follows:

$$\int_0^\infty f(t)g(t)dt = \frac{1}{2\pi} \int_{-\infty}^{+\infty} f^*(i\omega)g(i\omega)d\omega \,.$$

The star in this equation denotes the complex conjugation.

---

[8] Some authors refer to this property as the Parseval identity. However, according to the *Mathematical Encyclopedia* [140] and the book by Kolmogorov and Fomin [70], the attribution to Plancherel is correct.

If in the above discussion the functions are components of a vector or a matrix, then all the products are defined in the appropriate vector space.

A.6   Quadratic and Hermitian Forms

Given a real symmetric $n \times n$ matrix A, the quadratic form for a vector $x \in \mathbb{R}^n$ is the expression $\mathcal{F}(x) = x^* A x$. Note that there is no loss of generality in requiring the matrix A to be symmetric, because if it is not, the identical expression is obtained by replacing it with the symmetric matrix $(A^* + A)/2$.

Consider now the case with $x \in \mathbb{C}^n$ and let A be an $n \times n$ matrix with complex components. If this matrix is Hermitian, that is, $A^* = A$ (Hereafter the star will denote the transpose operation followed by taking the conjugate of the complex number. For real vectors or matrices, this reduces to just the transpose operation), the expression $\mathcal{F}(x) = x^* A x$ is real and is called a Hermitian form. In a general case, the Hermitian form is the expression $\mathcal{F}(x) = \operatorname{Re} x^* A x$. In other words, the value of the Hermitian form is always a real number.

If this value is positive (respectively, nonnegative, nonpositive, negative) for all $x \neq 0$, the form is called positive-definite (respectively, positive-semidefinite, negative-semidefinite, negative-definite). We call the symmetric (or Hermitian) matrix positive-definite, and write $A > 0$, if the corresponding quadratic (or Hermitian) form is positive-definite. In a similar manner we define positive-semidefinite, negative-semidefinite, and negative-definite symmetric or Hermitian matrices.

Every quadratic form can be extended to a Hermitian form by formally allowing the components of the real vector to be complex and taking the real part of the resulting expression. This process and the resulting expression will be called a Hermitian extension of a quadratic form.

Furthermore, for any Hermitian form there exists a Hermitian matrix B, such that $\operatorname{Re} x^* A x = x^* B x$. It is not difficult to find that this matrix B is given by

$$B = \frac{1}{2}(A + A^*).$$

The expression in the right-hand side will be called the real part of the matrix A and denoted $\operatorname{Re} A$. From now on, when we write this, we will not be concerned whether the matrix A is real or complex and use the last expression as a formal definition of $\operatorname{Re} A$.

Let us illustrate these concepts with a simple example. Consider the quadratic form:

$$\mathcal{F}(x) = x_1(x_2 - 3x_1).$$

In the vector-matrix notation $x^* A x$, we have

$$A = \begin{bmatrix} -3 & 1 \\ 0 & 0 \end{bmatrix}$$

The corresponding Hermitian form $\mathrm{Re}\, x^* A x = x^* B x$ will have the matrix:

$$B = \mathrm{Re}\, A = \frac{1}{2}(A + A^*) = \begin{bmatrix} -3 & 1/2 \\ 1/2 & 0 \end{bmatrix}.$$

If the components of the vector x are functions of a real variable t, then by Plancherel theorem we have

$$\int_0^\infty \mathcal{F}(x(t))dt = \frac{1}{4\pi^2} \int_{-\infty}^{+\infty} \tilde{\mathcal{F}}(\hat{x}(i\omega))d\omega,$$

where the tilde denotes the Hermitian extension of the quadratic form. This equality means that the Fourier transform preserves the sign-definite property of the quadratic form.

References

1. Aggrawal, T., Salapaka, M.V.: Real-Time Nonlinear Correction of Back-Focal-Plane Detection in Optical Tweezers. Rev. Sci. Inst. 81, 123105 (2010)
2. Aizerman, M.A.: On a Problem Concerning Stability "in the Large" of Dynamical Systems. Uspekhi Matematicheskikh Nauk 4, 187–188 (1949) (in Russian)
3. Aizerman, M.A., Gantmacher, F.R.: Absolute Stability of Regulator Systems. Holden-Day, San Francisco (1964)
4. Altshuller, D.A.: Zames-Falb Multipliers for Systems with Time-Periodic Nonlinearities. In: Proc. 2002 American Control Conf. (2002)
5. Altshuller, D.A.: A Generalization of the Frequency-Domain Stability Criteria to a Wider Class of Systems. In: Proc. 2002 Conf. on Decision and Control (2002)
6. Altshuller, D.A.: Stability Multipliers for Systems with Nonstationary Nonlinearities. Vestnik of the St. Petersburg State University, 3–12 (2003) (in Russian)
7. Altshuller, D.A.: Absolute Stability of Control Systems with Nonstationary Nonlinearities. Ph.D. Dissertation, St. Petersburg State University, Russia (2004) (in Russian)
8. Altshuller, D.A.: Frequency-Domain Stability Criteria for Two Classes of Systems with Time-Periodic Feedback. In: Proc. 2004 Symp. on Systems, Structure, and Control (2004)
9. Altshuller, D.A.: Frequency-Domain Criterion for Stability of Oscillations in a Class of Nonlinear Feedback Systems. In: Proc. 2007 Conf. on Periodic Control Systems, St. Petersburg, Russia (2007)
10. Altshuller, D.A.: A Partial Solution of the Aizerman Problem for Second-Order Systems with Delays. IEEE Trans. Aut. Control 53, 2058–2060 (2008)
11. Altshuller, D.A.: Delay-Integral-Quadratic Constraints and Absolute Stability of Time-Periodic Feedback Systems. SIAM J. Control Optim. 47, 3185–3202 (2009)
12. Altshuller, D.A.: Three Classes of Time-Delay Systems, for which the Aizerman Conjecture is True (2009),
http://press.princeton.edu/math/blondel/
problem_6_6_cont.pdf
13. Altshuller, D.A.: Frequency-Domain Criteria for Robust Stability for a Class of Linear Time-Periodic Systems. In: Proc. 2010 American Control Conf. (2010)
14. Altshuller, D.A.: Delay-Integral-Quadratic Constraints and Stability Multipliers for Systems with MIMO Nonlinearities. IEEE Trans. Aut. Control 56, 738–747 (2011)
15. Altshuller, D.A., Proskurnikov, A.P., Yakubovich, V.A.: Frequency-Domain Criteria for Dichotomy and Absolute Stability for Integral Equations with Quadratic Constraints Involving Delays. Doklady Math. 70, 998–1002 (2004)
16. Anderson, B.D.O., Vongpanitlerd, S.: Network Analysis and Synthesis. Dover, Mineola (2006)

17. Barabanov, N.E.: A Dichotomy of Nonlinear Control Systems that Satisfy a Differential Constraint. Automation and Remote Control 43, 137–139 (1982)
18. Barabanov, N.E.: Stability, Instability, and Dichotomy of Control Systems with Gradient Nonlinearities. Automation and Remote Control 43, 425–428 (1982)
19. Barabanov, N.E.: Frequency Criteria for Stability and Instability in the Large of the Stationary Sets of Nonlinear Systems of Differential Equations with a Single Monotone Nonlinearity. Siberian Math. J. 28, 191–202 (1987)
20. Barabanov, N.E.: New Criteria for Absolute Stability for Control Systems with One Differentiable Nonlinearity. Doklady Akademii Nauk 299, 570–572 (1988) (in Russian)
21. Barabanov, N.E.: On Kalman Problem. Siberian Math. J. 29, 333–341 (1988)
22. Barabanov, N.E.: New Frequency Criteria for Absolute Stability and Instability of Automatic-Control Systems with Nonstationary Nonlinearities. Differential Equations 25, 367–373 (1989)
23. Barabanov, N.E.: On the Aizerman Problem for Third-Order Nonstationary Systems. Differential Equations 29, 1439–1448 (1993)
24. Barabanov, N.E.: The State Space Extension Method in the Theory of Absolute Stability. IEEE Trans. Aut. Control 45, 2335–2339 (2000)
25. Barabanov, N.E., Gelig, A.K., Leonov, G.A., Likhtarnikov, A.L., Matveev, A.S., Smirnova, V.B., Fradkov, A.L.: The Frequency Theorem (Kalman-Yakubovich Lemma) in Control Theory. Automation and Remote Control 57, 1377–1407 (1996)
26. Barabanov, N.E., Yakubovich, V.A.: Absolute Stability of Control Systems with One Hysteresis Nonlinearity. Automation and Remote Control 40, 1713–1719 (1979)
27. Barkin, A.I.: L_{2p}-Stability and Absolute Stability of Nonlinear Systems. Automation and Remote Control 44, 1290–1295 (1983)
28. Barkin, A.I.: The Relationship between Two Absolute Stability Criteria. Automation and Remote Control 45, 29–34 (1984)
29. Barkin, A.I., Zelentsovskii, A.L.: Absolute Stability Criterion for Nonli-near Control Systems. Automation and Remote Control 42, 853–857 (1981)
30. Basar, T. (ed.): Control Theory: Twenty-Five Seminal Papers (1932-1981). IEEE Press, New York (2001)
31. Bergen, A.R., Sapiro, M.A.: The Parabola Test for Absolute Stability. IEEE Trans. Aut. Control AC-12, 312–314 (1967)
32. Benton, E.R.: Supersonic Magnus Effect on a Finned Missile. AIAA Journal 2, 153–155 (1964)
33. Boyd, S., El Ghaoui, L., Feron, E., Balakrishnan, V.: Linear Matrix Inequalities in Systems and Control Theory. SIAM, Philadelphia (1994)
34. Brockett, R.W., Willems, J.C.: Frequency-Domain Stability Criteria – Part I. IEEE Trans. Aut. Control AC 10, 255–261 (1965)
35. Brockett, R.W., Willems, J.C.: Frequency-Domain Stability Criteria – Part II. IEEE Trans. Aut. Control AC 10, 407–413 (1965)
36. Bongiorno Jr., J.J.: An Extension of the Nyquist-Barkhausen Stability Criterion to Linear Lumped-Parameter Systems with Time-Varying Elements. IEEE Trans. Aut. Control AC-8, 166–170 (1963)
37. Chang, M., Mancera, R., Safonov, M.: Computation of Zames-Falb Multipliers Revisited. In: Proc. 49th IEEE Conf. on Decision and Control (2010)
38. Chang, M., Mancera, R., Safonov, M.: Computation of Zames-Falb Multipliers Revisited. IEEE Trans. Aut. Control 57, 1024–1029 (2012)

39. Cho, Y.S., Narendra, K.S.: An Off-Axis Circle Criterion for the Stability of Feedback Systems with a Monotone Nonlinearity. IEEE Trans. Aut. Control AC-13, 413–416 (1968)
40. Corduneanu, C.: Integral Equations and Stability of Feedback Systems. Academic Press, New York (1973)
41. Coppel, W.A.: Dichotomies in Stability Theory. Springer (1978)
42. Dahleh, M., Tesi, A., Vicino, A.: On the Robust Popov Criterion for Interval Lur'e System. IEEE Trans. Aut. Control 38, 1400–1405 (1993)
43. De Figueiredo, R.J.P., Chen, G.: Nonlinear Feedback Control Systems: An Operator Theory Approach. Academic, San Diego (1993)
44. Desoer, C.A., Vidyasagar, M.: Feedback Systems: Input-Output Properties. Academic, New York (1975)
45. Dewey, A.G., Jury, E.I.: A Stability Inequality for a Class of Nonlinear Feedback Systems. IEEE Trans. Aut. Control AC-11, 54–62 (1996)
46. El'sgol'ts, L.E., Norkin, S.B.: Introduction to the Theory and Applica-tions of Differential Equations with Deviating Arguments. Academic, New York (1973)
47. Falb, P.L., Freedman, M.I.: A Generalized Transform Theory for Causal Operators. SIAM J. Control 7, 452–471 (1969)
48. Fitts, R.E.: Two Counterexamples to Aizerman's Conjecture. IEEE Trans. Aut. Control AC-11, 553–556 (1966)
49. Freedman, M.I.: Phase Function Norm Estimates for Stability of Systems with Monotone Nonlinearities. SIAM J. Control 10, 99–111 (1972)
50. Freedman, M.I., Falb, P.L., Zames, G.: Hilbert Space Stability Theory over Locally Compact Abelian Groups. SIAM J. Control 7, 479–495 (1969)
51. Gapski, P.B., Geromel, J.C.: A Convex Approach to the Absolute Stabil-ity Problem. IEEE Trans. Aut. Control 39, 1929–1932 (1994)
52. Gelig, A.K., Churilov, A.N.: Stability and Oscillations of Nonlinear Pulse-Modulated Systems. Birkhauser, Boston (1998)
53. Gil', M.I.: Stability of Finite and Infinite Dimensional Systems. Kluwer Academic, Boston (1998)
54. Gil', M.: Explicit Stability Conditions for Continuous Systems: A Functional Analysis Approach. Springer, Berlin (2005)
55. Grujic, L.: On Absolute Stability and Aizerman Conjecture. Automatica 17, 335–349 (1981)
56. Gusev, S.V., Likhtarnikov, A.L.: Kalman-Popov-Yakubovich Lemma and the S-Procedure: A Historical Essay. Automation and Remote Control 67, 1768–1810 (2006)
57. Haddad, W., Kapila, V.: Absolute Stability Criteria for Multiple Slope-Restricted Monotonic Nonlinearities. IEEE Trans. Aut. Control 40, 361–365 (1995)
58. Hahn, W.: Stability of Motion. Springer, Berlin (1969)
59. Halanay, A.: Differential Equations: Stability, Oscillations, Time Lags. Academic, New York (1996)
60. Halanay, A., Rasvan, V.: Stability and Stable Oscillations in Discrete Time Systems. CRC, Amsterdam (2000)
61. Holtzman, J.M.: Nonlinear Systems Theory: A Functional Analysis Approach. Prentice Hall, New York (1970)
62. Jury, E.I.: Inners and Stability of Dynamic Systems. Wiley, New York (1974)
63. Kalman, R.E.: Physical and Mathematical Mechanisms of Instability in Nonlinear Automatic Control Systems. Trans. ASME 79, 553–566 (1957)

64. Kalman, R.E.: Lyapunov Functions for the Problem of Lur'e in Automatic Control. Proc. National Acad. Sci. 49, 201–205 (1963)
65. Kamenetskii, V.A.: Absolute Stability and Absolute Instability of Control Systems with Several Nonlinear Nonstationary Elements. Automation and Remote Control 44, 1543–1552 (1983)
66. Kamenetskii, V.A.: Convolution Method for Matrix Inequalities and Absolute Stability Criteria for Stationary Control Systems. Automation and Remote Control 50, 598–607 (1989)
67. Kamenetskii, V.A., Pyatnitskii, E.S.: Gradient Method of Constructing Lyapunov Functions in Problems of Absolute Stability. Automation and Remote Control 48, 1–9 (1987)
68. Khalil, H.K.: Nonlinear Systems, 2nd edn. Prentice Hall, Upper Saddle River (1996)
69. Kheifetz, M.Z., Gelig, A.K.: On a Certain Type of Insensitivity of Regulator Systems. Energomashinostroenie (1), 25–26 (1964) (in Russian)
70. Kolmogorov, A.N., Fomin, S.V.: Introductory Real Analysis. Prentice Hall, Englewood Cliffs (1970)
71. Korenevskii, D.G.: Stability of Dynamical Systems under Uncertain Perturbations of Parameters: Algebraic Criteria, Kiev (1989) (in Russian)
72. Krasovskii, A.A.: Control Systems Theory Handbook, Moscow (1987) (in Russian)
73. Kulkarni, V.V.: Multipliers for Memoryless Incrementally Positive MIMO Nonlinearities. Ph.D. Dissertation, University of Southern California (2001)
74. Kulkarni, V.V., Pao, L.Y., Safonov, M.G.: Positivity Preservation Properties of Rantzer Multipliers. IEEE Trans. Aut. Control 56, 190–194 (2011)
75. Lefschetz, S.: Stability of Nonlinear Control Systems. Academic, New York (1965)
76. Leonov, G.A.: On the Necessity of a Frequency Condition for Absolute Stability in the Critical Case of a Pair of Purely Imaginary Roots. Soviet. Math. Dokl. 11, 1016–1020 (1970)
77. Leonov, G.A.: Extension of Popov's Frequency Criterion for Nonstationary Systems. Automation and Remote Control 41, 1494–1499 (1980)
78. Leonov, G.A., Burkin, I.M., Shepelyavyi, A.I.: Frequency Methods in Oscillation Theory. Kluwer, Boston (1996)
79. Leonov, G.A., Ponomarenko, D.V., Smirnova, V.B.: Frequency-Domain Methods for Nonlinear Analysis. World Scientific, Singapore (1996)
80. Leonov, G.A., Reitmann, V., Smirnova, V.B.: Non-Local Methods for Pendulum-Like Feedback Systems. Teubner, Stuttgart (1992)
81. Leonov, G.A., Smirnova, V.B.: Mathematical Problems in Phase Synchronization, Nauka, St. Petersburg (2000) (in Russian)
82. Letov, A.M.: Stability of Nonlinear Control Systems. Princeton University Press, Princeton (1961)
83. Liao, X., Yu, P.: Absolute Stability of Nonlinear Control Systems. Springer, Berlin (2008)
84. Liberzon, M.R.: New Results on Absolute Stability of Nonstationary Controlled Systems (Survey). Automation and Remote Control 40, 1124–1140 (1979)
85. Liberzon, M.R.: On the Problem of Stability of Motion of a Rotating Axisymmetric Flying Vehicle. Mekhanika Tverdogo Tela (5), 9–13 (1988) (in Russian)
86. Liberzon, M.R.: On the Stability of Controlled Flight of an Axisymmetric Rotating Vehicle with Incomplete Model Information. In: Proc. 1999 American Control Conference (1999)

87. Liberzon, M.R.: Essays on the Absolute Stability Theory. Automation and Remote Control 67, 1610–1934 (2006)
88. Lipatov, A.V.: Stability of Continuous Systems with One Nonlinearity. Doklady Akademii Nauk 260, 812–817 (1981) (in Russian)
89. Lipatov, A.V.: Graphical-Analytical Method of Verification of Stability of a Continuous System with One Monotone Nonlinearity in Case when Voronov's Criterion is Not Applicable. Doklady Akademii Nauk 267, 1069–1072 (1982) (in Russian)
90. Lipatov, A.V.: Stability of Stationary System with One Nonlinear Unit. I. Fundamental Theorems. Automation and Remote Control 43, 737–746 (1982)
91. Lipatov, A.V.: Stability of Stationary System with One Nonlinear Unit. II. Geometrical Criterion. Automation and Remote Control 43, 865–871 (1982)
92. Lipatov, A.V.: Graphical Stability Criteria for Continuous Systems with One Differentiable Nonlinearity. Automation and Remote Control 45, 323–331 (1984)
93. Lipatov, A.V., Sadykov, F.R., Soloveichik, G.Y.: Graphical Methods for the Stability Analysis of Continuous Systems with One Nonlinearity of Various Classes. Automation and Remote Control 46, 300–306 (1985)
94. Lurye, A.I., Postnikov, V.N.: A Contribution to the Theory of Stability of Control Systems. Applied Mathematics and Mechanics 8, 246–248 (1944)
95. Lyapunov, A.M.: The General Problem of the Stability of Motion, Fuller, A.T.(trans. and ed.). Taylor and Francis, London (1992)
96. Materassi, D., Baschieri, B., Tiribilli, B., Zuccheri, G., Samori, B.: An Open Source/Real-Time Atomic Force Microscope Architecture to Perform Customizable Force Spectroscopy. Rev. Sci. Instruments 80, 084301 (2009)
97. Materassi, D., Salapaka, M.V.: A Generalized Zames-Falb Multiplier. IEEE Trans. Aut. Control 56, 1432–1436 (2011)
98. Matrosov, V.M.: Method of Lyapunov Vector Functions in Feedback Systems. Automation and Remote Control 31, 1458–1468 (1972)
99. McShane, E.J., Botts, T.A.: Real Analysis. D. Van Nostrand, Inc., Princeton (1959)
100. Megretski, A.: Necessary and Sufficient Conditions for Stability. IEEE Trans. Aut. Control 38, 753–756 (1993)
101. Megretski, A., Rantzer, A.: System Analysis Via Integral Quadratic Constraints. IEEE Trans. Aut. Control 42, 819–830 (1997)
102. Miller, R.K.: Nonlinear Volterra Integral Equations. W.A. Benjamin, Menlo Park (1971)
103. Molchanov, A.P., Pyatnitskii, E.S.: Lyapunov Functions that Specify Necessary and Sufficient Conditions for Absolute Stability of Nonlinear Nonstationary Control Systems I. Automation and Remote Control 47, 344–354 (1986)
104. Molchanov, A.P., Pyatnitskii, E.S.: Lyapunov Functions that Specify Necessary and Sufficient Conditions for Absolute Stability of Nonlinear Nonstationary Control Systems II. Automation and Remote Control 47, 443–451 (1986)
105. Molchanov, A.P., Pyatnitskii, E.S.: Lyapunov Functions that Specify Necessary and Sufficient Conditions for Absolute Stability of Nonlinear Nonstationary Control Systems III. Automation and Remote Control 47, 620–630 (1986)
106. Narendra, K.S., Cho, Y.S.: Stability of Feedback Systems Containing a Single Odd Monotonic Nonlinearity. IEEE Trans. Aut. Control AC-12, 448–450 (1967)
107. Narendra, K.S., Goldwyn, R.M.: A Geometrical Criterion for the Stability of Certain Nonlinear Autonomous Systems. IEEE Trans. Circuit Theory CT-11, 406–408 (1964)
108. Narendra, K.S., Neuman, C.P.: Stability of a Class of Differential Equations with a Single Monotone Nonlinearity. SIAM J. Control 4, 295–308 (1966)

109. Narendra, K.S., Taylor, J.H.: Frequency Domain Criteria for Absolute Stability. Academic, New York (1973)

110. Naumov, B.N., Barkin, A.I., Sinitsyn, I.N.: Nonlinear Control Systems: Frequency Analysis of Absolute Stability and Performance. In: Naumov, B.N. (ed.) Philosophy of Nonlinear Control Systems. CRC, Boca Raton (1990)

111. Niculescu, S.-I.: Delay Effects on Stability. Springer, Berlin (2001)

112. O'Shea, R.P.: A Combined Frequency-Time Domain Stability Criterion for Autonomous Continuous Systems. IEEE Trans. Aut. Control AC-11, 477–484 (1996)

113. O'Shea, R.P.: An Improved Frequency Time Domain Stability Criterion for Autonomous Continuous Systems. IEEE Trans. Aut. Control AC-11, 725–731 (1996)

114. Pavlov, A., van der Wouw, N., Nijmeijer, H.: Uniform Output Regulation of Nonlinear Systems: A Convergent Dynamics Approach. Birkhauser, Boston (2006)

115. Pliss, V.A.: Certain Problems in the Theory of Stability of Motion in the Whole. NASA (1958)

116. Pontryagin, L.S.: On the Zeros of Some Elementary Transcendental Functions. In: Pontryagin, L.S. (ed.) Selected Scholarly Works, vol. 2. Nauka, Moscow (1988) (in Russian)

117. Popov, V.M.: Absolute Stability of Nonlinear Automatic Control Systems. Automation and Remote Control 22, 857–875 (1962); Reprinted in: Basar, T. (ed.): Control Theory: Twenty Five Seminal Papers. IEEE, New York (2001)

118. Popov, V.M.: Hyperstability of Control Systems. Springer, Berlin (1973)

119. Pyatnitskii, E.S.: New Research on the Absolute Stability of Automatic Control Systems (Review). Automation and Remote Control 29, 855–881 (1968)

120. Pyatnitskii, E.S.: Existence of Absolutely Stable Systems, for which the Popov Criterion Fails. Automation and Remote Control 33, 22–29 (1973)

121. Pyatnitskii, E.S., Skorodinskii, V.I.: Numerical Methods of Lyapunov Function Construction and their Application to the Absolute Stability Problem. Systems and Control Letters 2, 130–135 (1982)

122. Pyatnitskii, E.S., Skorodinskii, V.I.: Numerical Methods of Constructing the Lyapunov Functions and Absolute Stability Criteria in the Form of Numerical Procedures. Automation and Remote Control 44, 1427–1437 (1983)

123. Rantzer, A.: Friction Analysis Based on Integral Quadratic Constraints. Int. J. Robust and Nonlinear Control 11, 645–652 (2001)

124. Rasvan, V.: Absolute Stability of Automatic Control Systems with Delays. Nauka, Moscow (1983) (in Russian)

125. Rasvan, V.: Delay Dependent and Delay Independent Aizerman Problem. In: Blondel, V.D., Megretski, A. (eds.) Unsolved Problems in Mathematical Systems and Control Theory. Princeton University Press, Princeton (2004)

126. Rekasius, Z.V., Rowland, J.R.: A Stability Criterion for Feedback Systems Containing a Single Time-Varying Nonlinear Element. IEEE Trans. Aut. Control AC-10, 352–354 (1965)

127. Safonov, M.G., Wyetzner, G.: Computer-Aided Analysis Renders Popov Criterion Obsolete. IEEE Trans. Aut. Control AC-32, 1128–1131 (1987)

128. Safonov, M.G., Kulkarni, V.V.: Zames-Falb Multipliers for MIMO non-linearities. Internat. J. Robust Nonlinear Control 10, 1025–1103 (2000)

129. Schilling, R.L.: Measures, Integrals, and Martingales. Cambridge University Press, Cambridge (2005)

130. Siljak, D.D.: Nonlinear Systems: The Parameter Analysis and Design. Wiley, New York (1969)

131. Skorodinskii, V.I.: Absolute Stability of Nonstationary Motions of an Axisymmetrical Rotating Flying Vehicle. Mekhanika Tverdogo Tela (3), 17–21 (1984) (in Russian)
132. Skorodinskii, V.I.: Iterative Methods of Testing Frequency-Domain Criteria of Absolute Stability in Continuous-Time Control Systems I. Automation and Remote Control 52, 941–946 (1991)
133. Skorodinskii, V.I.: Iterative Methods for Checking Frequency-Domain Criteria of Absolute Stability in Continuous Control Systems II. Automation and Remote Control 52, 1082–1088 (1991)
134. Soshnikov, V.N., Fedorova, N.V.: On a Stationary Motion of an Axi-symmetrical Rotating Flying Vehicle. Mekhanika Tverdogo Tela (1), 170–175 (1974) (in Russian)
135. Sundareshan, M.K., Thathachar, M.A.L.: Construction of Stability Multipliers for Nonlinear Feedback Systems. Int. J. Systems Sci. 5, 277–285 (1974)
136. Turner, M.C., Kerr, M., Postlethwaite, I., Sofrony, J.: L2-gain Bounds for Systems with Slope-Restricted Nonlinearities. In: Proc. 2010 American Control Conf. (2010)
137. Venkatesh, Y.: Noncausal Multipliers for Nonlinear System Stability. IEEE Trans. Aut. Control AC-15, 195–204 (1970)
138. Venkatesh, Y.: Energy Methods in Time-Varying Systems Stability and Instability Analysis. Springer, Berlin (1977)
139. Vidyasagar, M.: Nonlinear Systems Analysis, 2nd edn. Prentice Hall, Englewood Cliffs (1993)
140. Vinogradov, I.M. (ed.): Mathematical Encyclopedia, vol. 4. Soviet Encyclopedia, Moscow (1984)
141. Voronov, A.A.: Absolutely Stable Systems with a Differentiable Non-decreasing Nonlinearity. Automation and Remote Control 39, 947–958 (1978)
142. Voronov, A.A.: Present State and Problems of Stability Theory. Automation and Remote Control 43, 573–592 (1982)
143. Voronov, A.A.: Basic Principles of Automatic Control Theory. Mir, Moscow (1985)
144. Vukic, Z., Kuljaca, L., Donlagic, D., Tesnjak, S.: Nonlinear Control Systems. Marcel Dekker, New York (2003)
145. Willems, J.C.: The Analysis of Feedback Systems. MIT, Cambridge (1971)
146. Willems, J.C., Gruber, M.: Comments on "A Combined Frequency-Time Stability Criterion for Autonomous Continuous Systems. IEEE Trans. Aut. Control AC-12, 217–219 (1967)
147. Yakubovich, V.A.: The Solution of Certain Matrix Inequalities in Automatic Control Theory. Soviet Mathematics (by American Math. Society), 620–623 (1962); Reprinted in Basar, T. (ed.): Control Theory: Twenty-Five Seminal Papers. IEEE, New York (2001)
148. Yakubovich, V.A.: Frequency-Domain Conditions for Absolute Stability of Nonlinear Automatic Control Systems. In: Proc. Interuniversity Conf. on Applied Stability Theory and Analytical Mechanics, Kazan Aviation Institute (1962) (in Russian)
149. Yakubovich, V.A.: Method of Matrix Inequalities in Stability Theory of Nonlinear Control Systems I. Absolute Stability of Forced Oscillations. Automation and Remote Control 25, 905–917 (1964)
150. Yakubovich, V.A.: Method of Matrix Inequalities in Stability Theory of Nonlinear Control Systems II. Absolute Stability in the Class of Nonlinearities with a Condition Involving the Derivative. Automation and Remote Control 26, 577–592 (1965)

151. Yakubovich, V.A.: Method of Matrix Inequalities in Stability Theory of Nonlinear Control Systems III. Absolute Stability of Systems with Hysteresis Nonlinearities. Automation and Remote Control 26, 753–763 (1965)

152. Yakubovich, V.A.: Frequency-Domain Conditions for Absolute Stability of Equilibriums and Forced Oscillations in Nonlinear Control Systems. In: Proc. International Conference on Multivariable and Discrete Automatic Control Systems, Prague (1965) (in Russian)

153. Yakubovich, V.A.: Frequency-Domain Conditions of Stability of Solutions of Nonlinear Integral Equations of Automatic Control. Vestnik of Leningrad Univ. Mathematics Mechanics Astronomy, 109–125 (1967) (in Russian)

154. Yakubovich, V.A.: Frequency Conditions for the Absolute Stability of Control Systems with Several Nonlinear or Linear Nonstationary Blocks. Automation and Remote Control 28, 857–880 (1967)

155. Yakubovich, V.A.: Absolute Stability of Pulsed Systems with Several Nonlinear or Linear but Nonstationary Blocks. Automation and Remote Control 28, 1301–1313 (1967)

156. Yakubovich, V.A.: Absolute Stability of Pulse Systems with Several Nonlinear or Linear Nonstationary Blocks. II. Automation and Remote Control 29, 244–263 (1968)

157. Yakubovich, V.A.: Absolute Stability of Nonlinear Control Systems: I. General Frequency Criteria. Automation and Remote Control 30, 1903–1912 (1970)

158. Yakubovich, V.A.: Absolute Instability of Nonlinear Control Systems: II. Systems with Nonstationary Nonlinearities. The Circle Criterion. Automation and Remote Control 31, 876–884 (1971)

159. Yakubovich, V.A.: S-Procedure in Nonlinear Control Theory. Vestnik of Leningrad Univ. Mathematics Mechanics Astronomy, 62–77 (1971) (in Russian)

160. Yakubovich, V.A.: Methods of the Theory of Absolute Stability. In: Nelepin, R.A. (ed.) Methods of Investigating of Nonlinear Automatic Control Systems. Nauka, Moscow (1975) (in Russian)

161. Yakubovich, V.A.: Contribution to the Abstract Theory of Absolute Stability of Nonlinear Systems. Vestnik of Leningrad Univ. Mathematics Mechanics Astronomy, 99–118 (1977) (in Russian)

162. Yakubovich, V.A.: Frequency Methods of Qualitative Investigations into Nonlinear Systems. In: Ishlinsky, A.Y., Chernousko, F.L. (eds.) Advances in Theoretical and Applied Mechanics. Mir, Moscow (1982)

163. Yakubovich, V.A.: Quadratic Criterion for Absolute Stability. Doklady Math. 58, 169–171 (1998)

164. Yakubovich, V.A.: Frequency-Domain Conditions for Stability of Nonlinear Systems. In: Matrosov, V.M., Vasilyeva, S.N., Moskalenko, A.I. (eds.) Nonlinear Control Theory and Its Applications. Nauka, Moscow (2000) (in Russian)

165. Yakubovich, V.A.: Necessity in Quadratic Criterion for Absolute Stability. Int. J. Robust Nonlinear Control 10, 889–907 (2000)

166. Yakubovich, V.A.: Popov's Method and Its Subsequent Development. European J. Control 8, 200–209 (2002)

167. Yakubovich, V.A., Leonov, G.A., Gelig, A.K.: Stability of Stationary Sets in Control Systems with Discontinuous Nonlinearities. World Scientific (2004)

168. Yakubovich, V.A., Starzhinskii, V.M.: Linear Differential Equations with Periodic Coefficients. Wiley, New York (1975)

169. Yakubovich, V.A., Yakubovich, D.V.: Local Analogue of the Frequency-Domain Popov Criterion for Stability of a Nonlinear System. Doklady Akademii Nauk. 371, 462–465 (2000) (in Russian)
170. Zames, G.: On the Input-Output Stability of Time-Varying Feedback Systems. Part I: Conditions Derived Using Concepts of Loop Gain, Conicity, and Positivity. IEEE Trans. Aut. Control AC-11, 228–238 (1966)
171. Zames, G.: On the Input-Output Stability of Time-Varying Feedback Systems. Part II: Conditions Involving Circles in the Frequency Plane and Sector Nonlinearities. IEEE Trans. Aut. Control AC-11, 465–476 (1966)
172. Zames, G., Falb, P.L.: On the Stability of Systems with Monotone and Odd Monotone Nonlinearities. IEEE Trans. Aut. Control AC-12, 221–223 (1967)
173. Zames, G., Falb, P.L.: On the Stability of Systems with Monotone and Slope-Restricted Nonlinearities. SIAM J. Control 6, 89–108 (1968)
174. Zames, G., Kallman, R.R.: On Spectral Mappings, Higher Order Circle Criteria, and Periodically Varying Systems. IEEE Trans. Aut. Control AC-15, 649–652 (1970)
175. Zhermolenko, V.N., Lokshin, B.Y.: On the Stability of Nonstationary Motions of an Axisymmetrical Rotating Flying Vehicle. Mekhanika Tverdogo Tela (5), 32–39 (1977)

Index